AF595130

Mycobacterium Tuberculosis Pathogenesis, Infection Prevention and Treatment

Mycobacterium Tuberculosis Pathogenesis, Infection Prevention and Treatment

Editors

Riccardo Miggiano
Menico Rizzi
Davide M. Ferraris

MDPI • Basel • Beijing • Wuhan • Barcelona • Belgrade • Manchester • Tokyo • Cluj • Tianjin

Editors

Riccardo Miggiano
University of Eastern Piedmont
Italy

Menico Rizzi
University of Eastern Piedmont
Italy

Davide M. Ferraris
University of Eastern Piedmont
Italy

Editorial Office
MDPI
St. Alban-Anlage 66
4052 Basel, Switzerland

This is a reprint of articles from the Special Issue published online in the open access journal *Pathogens* (ISSN 2076-0817) (available at: https://www.mdpi.com/journal/pathogens/special_issues/Mycobacterium_tuberculosis).

For citation purposes, cite each article independently as indicated on the article page online and as indicated below:

LastName, A.A.; LastName, B.B.; LastName, C.C. Article Title. *Journal Name* **Year**, *Article Number*, Page Range.

ISBN 978-3-03936-658-3 (Hbk)
ISBN 978-3-03936-659-0 (PDF)

Cover image courtesy of Riccardo Miggiano.

Contents

About the Editors

Riccardo Miggiano graduated in Pharmaceutical Chemistry and Technology at the University of Parma and received a Ph.D. in Pharmaceutical and Food Biotechnologies at University of Piemonte Orientale. By adopting different experimental approaches, including the use of complementary biochemical and biophysical techniques his research activity mainly focus on (i) the biochemical and structural characterization of individual enzymes involved in alkylated-DNA repair in Mycobacterium tuberculosis (Mtb); (ii) the identification and characterization of macromolecular complexes participating to the maintenance of Mtb genome stability and/or to the co-ordination of DNA repair with other vital aspects of the pathogen biology; (iii) protein engineering studies for the design of self-labeling fluorescent protein-tag. He was visiting researcher at the Institute of Biosciences and BioResources of CNR in Naples, at the Centre for Genomic Regulation in Barcelona and at the City University of New York. In 2017 Dr. Miggiano was awarded with a research grant from Fondazione Cariplo that sees him as Principal investigator (PI) and coordinator of the project "Deciphering molecular aspects of Mycobacterium tuberculosis DNA repair to disclose its role in the pathogenesis of tuberculosis in humans". His activity resulted in 23 published papers and 19 protein structures deposited in the Protein Data Bank. Moreover, Dr. Miggiano is a co-founder of the IXTAL spin-off (www.ixtal.it), a biotech company that operates in the field of protein science.

Menico Rizzi graduated in Chemistry at the University of Pavia Italy and received a Ph.D. in Molecular Biotechnologies at University Cattolica "Sacro Cuore". He is an expert in structural biology and enzymology and coordinates a group investigating proteins of medical interest, with a particular focus on poverty-linked diseases. Prof. Rizzi has a successful record as a group leader in European Commission funded projects and received funding also from National Agencies and Private Foundations. Visiting scientist/professor at the University of York (UK), University of Kyoto (JP) and at the Japanese National Institute for infectious disease (Tokyo, JP). Member of the scientific committee of several National and International Congresses including FEBS and FASEB meetings and lecturer in International PhD Schools supported by UNESCO. Past President of the "Proteins" division of the Italian Society of Biochemistry and Molecular Biology and of the Evaluation Board of the University of Piemonte Orientale and University of Genova (Italy). Expert member of the International nonproprietary name (INN) program of the World Health Organization. Since April 2020 Prof. Rizzi is a member of the governing board of ANVUR, the Italian National Agency for the evaluation of Universities and Research Institutes.

Davide M. Ferraris graduated in Chemistry at the University of Torino (Italy) and was a Marie-Curie fellow at the Helmholtz Centre for Infection Research (HZI; Braunschweig, Germany) and received a Ph.D. in Infection Biology from the Hannover Medical School (MHH, Germany). His interests and activities include the structural and biochemical studies of enzymes and proteins involved in infectious diseases, and the development of enzymes for custom industrial applications. He was the recipient of the 2017 "Roche for Research" grant, and is the co-founder and Managing Director of the company IXTAL (www.ixtal.it) which offers R&D services in protein sciences to pharmaceutical and biotech companies.

Editorial

Mycobacterium tuberculosis Pathogenesis, Infection Prevention and Treatment

Riccardo Miggiano, Menico Rizzi and Davide M. Ferraris *

Department of Pharmaceutical Sciences, University of Piemonte Orientale, Via Bovio 6, 28100 Novara, Italy; riccardo.miggiano@uniupo.it (R.M.); menico.rizzi@uniupo.it (M.R.)

* Correspondence: davide.ferraris@uniupo.it; Tel.: +39-0321375715

Received: 13 May 2020; Accepted: 15 May 2020; Published: 18 May 2020

Abstract: Tuberculosis (TB) is an infectious disease caused by the bacterium Mycobacterium tuberculosis (MTB) and it represents a persistent public health threat for a number of complex biological and sociological reasons. According to the most recent Global Tuberculosis Report (2019) edited by the World Health Organization (WHO), TB is considered the ninth cause of death worldwide and the leading cause of mortality by a single infectious agent, with the highest rate of infections and death toll rate mostly concentrated in developing and low-income countries. We present here the editorial section to the Special Issue entitled "Mycobacterium tuberculosis Pathogenesis, Infection Prevention and Treatment" that includes 7 research articles and a review. The scientific contributions included in the Special Issue mainly focus on the characterization of MTB strains emerging in TB endemic countries as well as on multiple mechanisms adopted by the bacteria to resist and to adapt to antitubercular therapies.

Keywords: *M. tuberculosis*; tuberculosis; host-pathogen interactions; immune response; antitubercular drug discovery; antitubercular treatments

---

Editorial

Tuberculosis (TB) is an infectious disease caused by the bacterium *Mycobacterium tuberculosis* (MTB) and it represents a persistent public health threat for a number of complex biological and sociological reasons. According to the most recent Global Tuberculosis Report (2019) edited by the World Health Organization (WHO) [1], TB is considered the ninth cause of death worldwide and the leading cause of mortality by a single infectious agent, with the highest rate of infections and death toll rate mostly concentrated in developing and low-income countries. TB is also considered an impairing factor for economic growth and for the improvement of the general public health in those countries, as it drains human and financial resources that would otherwise be invested in the economy [2]. Hence, there is a pressing need to study and develop new prevention protocols and treatments for TB. Public health policy makers, supranational organizations and governing bodies are currently joining efforts in raising awareness in the general population regarding MTB contagion and in establishing guidelines and protocols for fighting TB [3]. At the same time, pharmaceutical companies research new therapies and approaches for finding new antitubercular diagnostics and treatments, the commercial sustainability of which should not be overlooked in order to make antitubercular treatments accessible and inclusive [4].

Research articles and the review published in this Special Issue entitled "*Mycobacterium tuberculosis* Pathogenesis, Infection Prevention and Treatment" mainly focus on the characterization of MTB strains emerging in TB endemic countries as well as on multiple mechanisms adopted by the bacteria to resist and to adapt to antitubercular therapies. The work of Mupfumi L. et al. investigates the dynamics of the host immune response during MTB infection in HIV/TB co-infected patients, defining the functional, activation, and differentiation profile of MTB-specific T-cells during antiretroviral treatment [5]. The research article by Fursov et al. [6] investigates the genetic and phenotypic profile of the MTB strain Rostov, belonging to the Central Asia Outbreak Clade (CAO) of the Beijing genotype. This strain has been attributed to the pre-extensively drug-resistant (XDR) tuberculosis group. In particular, the authors analyzed the growth rate and virulence of the Rostov strain in mice models, and the experimental outputs were compared with the same characteristics of the MTB H37Rv strain. However, mice infected with the Rostov strain did not show the formation of pulmonary infiltrates, suggesting a lower activation of the host defense mechanisms compared with the response to the infection caused by H37Rv strain.

The emergence and global spread of multidrug-resistant (MDR) as well as XDR MTB strains requires the early detection of drug resistance to ensure a functional patient management. In this context, Mogashoa and co-authors contribute to the Special Issue by presenting an evaluation of the second line drug resistance among drug resistant MTB isolates in Botswana [7]. The study analyzed 57 clinical isolates demonstrating that 33 (58%) were MDR strains, 4 (7%) were additionally resistant to flouroquinolones, and 3 (5%) were resistant to both fluoroquinolones and second-line injectable drugs. Moreover, they detected the most conserved mutation conferring he resistance to fluoroquinolone treatments, located on the *gyrA* gene with the alanine at position 90 mutated into valine (A90V).

For a definitive solution to the clinical management of drug-resistant tuberculosis, other innovative drugs targeting alternative pathways are urgently needed taking into account the genes that are essential for growth and survival of the bacilli *in vitro* [8], in macrophages [9] and in animal models of infection [10]. Among these, alternative validated target pathways include DNA transcription, targeted by rifampicin, protein synthesis, which is inhibited by oxazolidinones [11] and ATP synthesis by Q203 [12] and bedaquiline [13]. Although the DNA metabolic pathway plays a key role in mutagenesis events conferring bacterial drug resistance, a limited number of approved TB drugs target DNA metabolism [14], which includes key enzymatic steps involved in nucleotides synthesis [14–18], DNA replication [19] and repair [20,21]. As described by Uddin R. and collaborators [22], innovative targets could be identified also by the computational subtractive genomics methods; indeed, the authors presented a prioritized list of possible targets for drug discovery studies against *Mycobacterium avium* sub. *hominissuis*. In addition to drug-resistance, research efforts should take into account alterations of the metabolic profile occurring to MTB bacilli during anti-tubercular treatments. To this end, the work of Bespyatykh et al. [23] demonstrates the occurrence of changes in bacterial metabolism during TB therapy using multi-omics analysis of three consecutive MTB isolates from the same patient. In particular, they observed a stepwise accumulation of polymorphisms related to phenotypic resistance to fluoroquinolones and isoniazid and variations at the proteomic and transcriptomic levels, in the *loci* associated with drug-resistance and virulence, that only partly can be explained by mutagenic events on target genes. In support of this hypothesis, the work of Maslov D.A. showed that mutations on a transcriptional regulator gene (MSMEG_1380) indirectly confer resistance to tetrazines by inducing the overexpression of the mmpS5-mmpL5 operon that regulates drug efflux in *Mycobacterium smegmatis* [24].

MTB pathogenicity is mainly based on (i) the capability of the bacilli of reprogramming host macrophages after primary infection, preventing its own elimination; (ii) the formation of granulomas, in which the pathogen survives in equilibrium with the host defense and (iii) the slowing control of bacterial central metabolism and replication, characterizing the so called dormant state in which MTB is resistant to host defenses and therapy. Since dormant bacilli could also reside in bone marrow mesenchymal stem cells, as observed in post-chemotherapy mice models and clinical subjects, the paper of Garhyan J. et al. [25] presents innovative bone-homing PEGylated liposome nanoparticles which actively target the bone microenvironment leading to MTB clearance and reducing the relapse rate. Concerning the macrophages reprogramming capability, Abdalla A.E. and co-authors contribute to the Special Issue with a review discussing the multiple mechanisms adopted by MTB to interfere with macrophage apoptosis [26]. In particular, they describe the anti-apoptotic determinants, listing the mechanism of action and the main molecular outcome. Moreover, the authors focus on the bacterial capability to selectively regulate both the release of anti-apoptotic cytokines and the expression of microRNAs whose up-regulation is related to apoptosis blocking.

The articles published in this Special Issue deal with different aspects of TB pathogenesis and reflect the complexity of the disease management that demands a multi-disciplinary approach aimed at the understanding of each step of the infection cycle. The scientific papers that contribute to the Special Issue represent a part of the research efforts engaged in fighting TB that include a huge area of investigation with thousands of active researchers working in many complementary directions.

Author Contributions: Conceptualization, R.M., M.R. and D.M.F; writing—original draft preparation, R.M., M.R. and D.M.F.; writing—review and editing, R.M., M.R. and D.M.F.; supervision, M.R.; project administration, D.M.F. All authors have read and agreed to the published version of the manuscript.

Funding: This research received no external funding.

Acknowledgments: The Authors together with Franca Rossi and Silvia Garavaglia, as staff scientists of the Biochemistry and Structural Biology Unit, owe a debt of gratitude to Castrese Morrone, Edoardo L.M. Gelardi, Eugenio Ferrario, Daniele Mazzoletti and Andrea Gatta for their contribution to ensure research activities during the difficult period of Covid-19 outbreak.

Conflicts of Interest: The authors declare no conflict of interest.

References

1. World Health Organization. *Global tuberculosis Report WHO 2018*; WHO: Geneva, Switzerland, 2018; Volume 69.
2. Reid, M.J.A.; Arinaminpathy, N.; Bloom, A.; Bloom, B.R.; Boehme, C.; Chaisson, R.; Chin, D.P.; Churchyard, G.; Cox, H.; Ditiu, L.; et al. Building a tuberculosis-free world: The Lancet Commission on tuberculosis. *Lancet* **2019**, *393*, 1331–1384. [CrossRef]
3. Political Declaration of the High-Level Meeting of the General Assembly on the Fight against Tuberculosis. 2018. Available online: https://digitallibrary.un.org/record/1649568 (accessed on 16 May 2020).
4. Treatment Action Group. *Tuberculosis Research Funding Trends 2005–2017*; Treatment Action Group: New York, NY, USA, 2018; Available online: https://www.treatmentactiongroup.org/wp-content/uploads/2018/11/tb_funding_2018_final.pdf (accessed on 16 May 2020).
5. Mupfumi, L.; Mpande, C.A.; Reid, T.; Moyo, S.; Shin, S.S.; Zetola, N.; Mogashoa, T.; Musonda, R.M.; Kasvosve, I.; Scriba, T.J.; et al. Immune Phenotype and Functionality of Mtb-Specific T-Cells in HIV/TB Co-infected patients on antiretroviral treatment. *Pathogens* **2020**, *9*, 180. [CrossRef] [PubMed]
6. Fursov, M.V.; Shitikov, E.A.; Bespyatykh, J.A.; Bogun, A.G.; Kislichkina, A.A.; Kombarova, T.I.; Rudnitskaya, T.I.; Grishenko, N.S.; Ganina, E.A.; Domotenko, L.V.; et al. Genotyping, assessment of virulence and antibacterial resistance of the rostov strain of mycobacterium tuberculosis attributed to the central asia outbreak clade. *Pathogens* **2020**, *9*, 335. [CrossRef] [PubMed]
7. Mogashoa, T.; Melamu, P.; Derendinger, B.; Ley, S.D.; Streicher, E.M.; Iketleng, T.; Mupfumi, L.; Mokomane, M.; Kgwaadira, B.; Rankgoane-Pono, G.; et al. Detection of second line drug resistance among drug resistant mycobacterium tuberculosis isolates in Botswana. *Pathogens* **2019**, *8*, 208. [CrossRef] [PubMed]
8. Sassetti, C.M.; Boyd, D.H.; Rubin, E.J. Genes required for mycobacterial growth defined by high density mutagenesis. *Mol. Microbiol.* **2003**, *48*, 77–84. [CrossRef]
9. Rengarajan, J.; Bloom, B.R.; Rubin, E.J. Genome-wide requirements for Mycobacterium tuberculosis adaptation and survival in macrophages. *Proc. Natl. Acad. Sci. USA* **2005**, *102*, 8327–8332. [CrossRef]
10. Sassetti, C.M.; Rubin, E.J. Genetic requirements for mycobacterial survival during infection. *Proc. Natl. Acad. Sci. USA* **2003**, *100*, 12989–12994. [CrossRef]
11. Jadhavar, P.S.; Vaja, M.D.; Dhameliya, T.M.; Chakraborti, A.K. Oxazolidinones as Anti-tubercular agents: Discovery, development and future perspectives. *Curr. Med. Chem.* **2015**, *22*, 4379–4397. [CrossRef]
12. Pethe, K.; Bifani, P.; Jang, J.; Kang, S.; Park, S.; Ahn, S.; Jiricek, J.; Jung, J.; Jeon, H.K.; Cechetto, J.; et al. Discovery of Q203, a potent clinical candidate for the treatment of tuberculosis. *Nat. Med.* **2013**, *19*, 1157–1160. [CrossRef]

13. Pym, A.S.; Diacon, A.H.; Tang, S.J.; Conradie, F.; Danilovits, M.; Chuchottaworn, C.; Vasilyeva, I.; Andries, K.; Bakare, N.; De Marez, T.; et al. Bedaquiline in the treatment of multidrug- and extensively drug-resistant tuberculosis. *Eur. Respir. J.* **2016**, *47*, 564–574. [CrossRef]
14. Miggiano, R.; Morrone, C.; Rossi, F.; Rizzi, M. Targeting genome integrity in mycobacterium tuberculosis: From nucleotide synthesis to DNA replication and repair. *Molecules* **2020**, *25*, 1205. [CrossRef]
15. Donini, S.; Garavaglia, S.; Ferraris, D.M.; Miggiano, R.; Mori, S.; Shibayama, K.; Rizzi, M. Biochemical and structural investigations on phosphoribosylpyrophosphate synthetase from Mycobacterium smegmatis. *PLoS ONE* **2017**, *12*, e0175815. [CrossRef] [PubMed]
16. Singh, V.; Donini, S.; Pacitto, A.; Sala, C.; Hartkoorn, R.C.; Dhar, N.; Keri, G.; Ascher, D.B.; Mondésert, G.; Vocat, A.; et al. The inosine monophosphate dehydrogenase, GuaB2, is a vulnerable new bactericidal drug target for tuberculosis. *ACS Infect. Dis.* **2017**, *3*, 5–17. [CrossRef] [PubMed]
17. Park, Y.; Pacitto, A.; Bayliss, T.; Cleghorn, L.A.; Wang, Z.; Hartman, T.; Arora, K.; Ioerger, T.R.; Sacchettini, J.; Rizzi, M.; et al. Essential but not vulnerable: Indazole sulfonamides targeting inosine monophosphate dehydrogenase as potential leads against mycobacterium tuberculosis. *ACS Infect Dis.* **2017**, *3*, 18–33. [CrossRef]
18. Donini, S.; Ferraris, D.M.; Miggiano, R.; Massarotti, A.; Rizzi, M. Structural investigations on orotate phosphoribosyltransferase from Mycobacterium tuberculosis, a key enzyme of the de novo pyrimidine biosynthesis. *Sci. Rep.* **2017**, *7*, 1180. [CrossRef] [PubMed]
19. Reiche, M.A.; Warner, D.F.; Mizrahi, V. Targeting DNA replication and repair for the development of novel therapeutics against tuberculosis. *Front. Mol. Biosci.* **2017**, *4*, 75. [CrossRef] [PubMed]
20. Ferraris, D.M.; Miggiano, R.; Rossi, F.; Rizzi, M. Mycobacterium tuberculosis molecular determinants of infection, survival strategies, and vulnerable targets. *Pathogens* **2018**, *7*, 17. [CrossRef]
21. Lahiri, S.; Rizzi, M.; Rossi, F.; Miggiano, R. Mycobacterium tuberculosis UvrB forms dimers in solution and interacts with UvrA in the absence of ligands. *Proteins* **2018**, *86*, 98–109. [CrossRef]
22. Uddin, R.; Siraj, B.; Rashid, M.; Khan, A.; Ahsan Halim, S.; Al-Harrasi, A. Genome subtraction and comparison for the identification of novel drug targets against mycobacterium avium subsp. hominissuis. *Pathogens* **2020**, *9*, 368. [CrossRef]
23. Bespyatykh, J.; Shitikov, E.; Bespiatykh, D.; Guliaev, A.; Klimina, K.; Veselovsky, V.; Arapidi, G.; Dogonadze, M.; Zhuravlev, V.; Ilina, E.; et al. Metabolic changes of mycobacterium tuberculosis during the anti-tuberculosis therapy. *Pathogens* **2020**, *9*, 131. [CrossRef]
24. Maslov, D.A.; Shur, K.V.; Vatlin, A.A.; Danilenko, V.N. MmpS5-MmpL5 Transporters Provide Mycobacterium smegmatis Resistance to imidazo[1,2-b][1,2,4,5]tetrazines. *Pathogens* **2020**, *9*, 166. [CrossRef] [PubMed]
25. Garhyan, J.; Mohan, S.; Rajendran, V.; Bhatnagar, R. Preclinical evidence of nanomedicine formulation to target mycobacterium tuberculosis at its bone marrow niche. *Pathogens* **2020**, *9*, 372. [CrossRef] [PubMed]
26. Abdalla, A.E.; Ejaz, H.; Mahjoob, M.O.; Alameen, A.A.M.; Abosalif, K.O.A.; Elamir, M.Y.M.; Mousa, M.A. Intelligent mechanisms of macrophage apoptosis subversion by mycobacterium. *Pathogens* **2020**, *9*, 218. [CrossRef] [PubMed]

Article

Detection of Second Line Drug Resistance among Drug Resistant Mycobacterium Tuberculosis Isolates in Botswana

Tuelo Mogashoa [1,2], Pinkie Melamu [2], Brigitta Derendinger [3], Serej D. Ley [3,4,5], Elizabeth M. Streicher [3], Thato Iketleng [2,6], Lucy Mupfumi [1,2], Margaret Mokomane [7], Botshelo Kgwaadira [8], Goabaone Rankgoane-Pono [8], Thusoyaone T. Tsholofelo [8], Ishmael Kasvosve [1], Sikhulile Moyo [2,9], Robin M. Warren [3] and Simani Gaseitsiwe [2,9,*]

1. Department of Medical Laboratory Sciences, Faculty of Health Sciences, University of Botswana, Gaborone 0022, Botswana; tuelomogashoa@me.com (T.M.); lmupfumi@gmail.com (L.M.); kasvosvei@ub.ac.bw (I.K.)
2. Botswana Harvard AIDS Institute Partnership, Gaborone 0000, Botswana; pmpmelamu@gmail.com (P.M.); iketlengt@gmail.com (T.I.); sikhulilemoyo@gmail.com (S.M.)
3. DST-NRF Centre of Excellence in Biomedical Tuberculosis Research, South African Medical Research Council Centre for Tuberculosis Research, Division of Molecular Biology and Human Genetics, Faculty of Medicine and Health Sciences, Stellenbosch University, Tygerberg, Cape Town 7505, South Africa; brigitta@sun.ac.za (B.D.); 21280606@sun.ac.za (S.D.L.); lizma@sun.ac.za (E.M.S.); rw1@sun.ac.za (R.M.W.)
4. Department of Medical Parasitology and Infection Biology, Swiss Tropical and Public Health Institute, Basel 4002, Switzerland
5. Faculty of Science, University of Basel, Basel 4001, Switzerland
6. College of Health Sciences, School of Laboratory Medicine and Medical Sciences, University of KwaZulu-Natal, Durban 4041, South Africa
7. National Tuberculosis Reference Laboratory, Ministry of Health and Wellness, Gaborone 0000, Botswana; bafanamargaret@gmail.com
8. Botswana National Tuberculosis Programme, Ministry of Health and Wellness, Gaborone 0000, Botswana; btkgwaadira@gmail.com (B.K.); goaba2000@yahoo.com (G.R.-P.); ttsholo808@gmail.com (T.T.T.)
9. Department of immunology and infectious diseases, Harvard T.H. Chan School of Public Health, Boston, Massachusetts, MA02115, USA

* Correspondence: sgaseitsiwe@bhp.org.bw

Received: 6 September 2019; Accepted: 23 October 2019; Published: 28 October 2019

Abstract: The emergence and transmission of multidrug resistant (MDR) and extensively drug resistant (XDR) *Mycobacterium tuberculosis (M.tb)* strains is a threat to global tuberculosis (TB) control. The early detection of drug resistance is critical for patient management. The aim of this study was to determine the proportion of isolates with additional second-line resistance among rifampicin and isoniazid resistant and MDR-TB isolates. A total of 66 *M.tb* isolates received at the National Tuberculosis Reference Laboratory between March 2012 and October 2013 with resistance to isoniazid, rifampicin or both were analyzed in this study. The genotypes of the *M.tb* isolates were determined by spoligotyping and second-line drug susceptibility testing was done using the Hain Genotype MTBDR*sl* line probe assay version 2.0. The treatment outcomes were defined according to the Botswana national and World Health Organization (WHO) guidelines. Of the 57 isolates analyzed, 33 (58%) were MDR-TB, 4 (7%) were additionally resistant to flouroquinolones and 3 (5%) were resistant to both fluoroquinolones and second-line injectable drugs. The most common fluoroquinolone resistance-conferring mutation detected was *gyrA* A90V. All XDR-TB cases remained smear or culture positive throughout the treatment. Our study findings indicate the importance of monitoring drug resistant TB cases to ensure rapid detection of second-line drug resistance.

Keywords: *Mycobacterium tuberculosis*; line probe assay; second-line drugs; drug resistance; XDR-TB; MDR-TB

1. Background

In 2017, 10 million people fell ill with tuberculosis (TB) and 1.6 million people died of TB [1]. In the same year, an increase in cases of rifampicin monoresistant and multidrug resistant TB (MDR-TB defined as TB that is resistant to rifampicin and isoniazid) from 490,000 in 2016 to 558,000 in 2017 was observed [1]. The increasing numbers of rifampicin and MDR-TB cases poses a risk to TB control programs throughout the world [2]. Rifampicin monoresistance is considered to be a precursor to MDR-TB and there are often concerns about rifampicin monoresistant TB patients acquiring MDR-TB [3,4]. The standardized World Health Organization (WHO) MDR-TB treatment regimen recommends the use of second-line injectable drugs (SLIDs) in combination with flouroquinolones as part of the standardized MDR-TB treatment regimen [2]. The resistance to a fluoroquinolone and a SLID negatively impacts treatment outcome and has been defined as extensively drug resistant TB (XDR-TB) [5–7]. MDR-TB in combination with resistance to either a fluoroquinolone or a SLID has been termed Pre-XDR-TB.

The resistance to fluoroquinolones is usually caused by point mutations in the quinolone resistance determining region (QRDR) of the gene encoding subunit A or B of the deoxyribonucleic acid (DNA) gyrase gene (*gyrA* or *gyrB*) [8–10]. In the *gyrA* gene, the resistance mutations are commonly found in codons 85 to 96, whereas for the *gyrB* gene, they are found in codons 472 and 510 [5,6]. The mutations between codon 1400 and 1500 in the *rrs* gene are often associated with resistance to SLIDs such as capreomycin, kanamycin and amikacin [5]. The timely detection of resistance to SLIDs remains critical for optimizing treatment to improve the treatment outcome as well as directing infection control measures to halt the transmission of drug resistant TB [7,11]. Diagnostic assays, such as the Hain GenoType MTBDR*sl* line probe assay (Hain Lifescience, Germany), have been endorsed by WHO for the rapid detection of second-line drug resistance [11]. The MTBDR*sl* test is based on the DNA strip technology which has three steps: DNA extraction, multiplex polymerase chain reaction (PCR) amplification and reverse hybridization [12]. This assay has proven to be reliable for rapidly detecting resistance to second-line drugs [13] and has been implemented in 28 countries in Africa [14]. Monitoring drug resistance with the help of such assays and evaluating treatment outcomes may help improve management of TB [15,16].

This study sought to determine the level of resistance to second-line drugs among rifampicin monoresistant, isoniazid monoresistant and multi-drug resistant TB cases in Botswana using the Hain genotype MTBDR*sl* Version 2.0 and to assess patient treatment outcomes.

2. Methods

2.1. Design and Study Population

This was a retrospective, cross-sectional study utilizing *M.tb* isolates from the Botswana National Tuberculosis Reference Laboratory (NTRL) bio-repository which were collected as part of routine clinical care between 2012 and 2013. The study was approved by the University of Botswana Ethics Institutional Review Board and Health Research and Development Committee (HRDC) at the Ministry of Health and Wellness (Reference No: HPDME: 13/18/1 Vol. XI (140)). Sixty-six *M.tb* isolates were selected. The selected isolates were isoniazid (H) or rifampicin (R) monoresistant or resistant to both (MDR) based on first-line culture-based drug susceptibility testing (DST). The isolates included in this study are part of a previously described larger study and culture-based drug susceptibility testing for first-line drugs was done as previously described [17]. The clinical treatment outcome data was obtained from the Botswana National Tuberculosis Program (BNTP) patient database. At the time of the study, the standardized MDR-TB treatment regimen in Botswana consisted of pyrazinamide, amikacin, levofloxacin, ethionamide, cycloserine, P-aminosalicylic acid (PAS) [18].

2.2. Treatment Outcome Definitions

Treatment outcomes were defined according to the Botswana national guidelines. Briefly, "cured" was defined as a patient whose smear or culture sample was positive at the start of treatment but either converted to smear negative or had two consecutive negative cultures, one during treatment and the other at the end of treatment; "failed" was defined as a patient whose smear or culture was positive five months or later during treatment; "loss-to-follow-up" was defined as a patient whose treatment was interrupted for more than 30 consecutive days; "not evaluated" referred to patients whose treatment outcome could not be assigned since treatment conclusion has not been reported to the national TB program; treatment "completed" referred to patients who completed treatment but did not have a negative smear or culture result in the last month of treatment [18]. In bivariate comparisons, treatment outcomes were combined: "failed treatment" (i.e. remaining smear or culture positive throughout treatment), "loss-to-follow-up", death, "not initiated on treatment" and "not evaluated" as "unsuccessful treatment outcome" and "completed treatment", "cured" as "successful treatment outcomes".

2.3. DNA Extraction

DNA was extracted from the BD MGIT960 cultures (BD Biosciences, Sparks, MD, USA) using the GenoLyse DNA extraction kit version 1.0 (Hain LifeScience, GmBH, Nehren, Germany) following the manufacturer's instructions [19].

2.4. Genotyping

2.4.1. Spoligotyping

The genotypes of the isolates were determined by spoligotyping as previously described by Kamerbeek et al. [20] and Mogashoa et al. [17]. The *M.tb* families and lineages of the isolates were assigned based on the spoligotyping results.

2.4.2. Hain Genotype MTBDRsl Version 2

The second line drug resistance profiles were determined by using the Hain GenoType MTBDR*sl* assay (Hain Lifescience, Germany). The steps were performed as per the manufacturer's instructions [12]. The culture based second line phenotypic DST was not performed for this study since the test was unavailable at the reference laboratory.

2.4.3. Data Analysis

Fischer's exact test was used to determine if there was an association between second line drug resistance and *M.tb* family, the patients' age, HIV status and sex. The factors were examined for a favorable treatment outcome using logistic regression techniques. A p-value of <0.05 was considered statistically significant. STATA version 14 (Stata Corp, College Station, TX, USA) was used for statistical analysis.

3. Results

A total of 57 out of 66 (86%) isolates (one isolate per patient) were successfully genotyped and tested for resistance to first-line drugs (culture-based phenotypic DST) and second-line drugs (line probe assay MTBDR*sl*). Of these 57 isolates, 27 (47.4%) were from the southern region, 24 (42.1%) were from the central region, 5 (8.8%) from north west and 1 (1.8%) from south west region. The median age of the patients was 34 years [Q_1, Q_3: 13,59] with half (50%) being in the 20–39 years age group. For those patients with a known HIV status, 31 (54.4%) were HIV positive, 15 (26.3%) were HIV negative and 11(19.3%) had an unknown HIV status. The *M.tb* lineages identified among the DR-TB isolates were Lineage 4 (66.7%), Lineage 2 (19.3%), Lineage 1 (12.3%) and unknown lineage (1.8%) (Table 1).

Table 1. Demographic characteristics of patients included in the study (n = 57).

	n	%
Sex		
Male	29	50.9
Female	28	49.1
	Age in years	
<20 years	9	16.1
20–39 years	28	50.0
40–59 years	16	28.6
>60 years	3	5.4
HIV status		
Negative	15	26.3
Positive	31	54.4
Unknown	11	19.3
	Specimen type	
Extra-pulmonary	1	1.8
Pulmonary	55	96.5
Unknown	1	1.8
Smear results		
Negative	12	21.1
Positive	45	79
	Drug resistance profile	
Rifampicin monoresistant	11	19.3
Isoniazid monoresistant	6	10.5
Multi-drug resistant (MDR)	33	57.9
Pre-XDR*	4	7.0
XDR**	3	5.3
	Region	
Central	24	42.1
South West	1	1.8
North West	5	8.8
Southern	27	47.4
	Lineage	
Lineage 1	7	12.3
Lineage 2	11	19.3
Lineage 4	38	66.7
Unknown	1	1.8

*Pre-XDR: Pre-extensively drug resistant. **XDR: extensively drug resistant.

Among the 57 drug resistant isolates, the first and second-line DST results showed that 19% of the cases were resistant to rifampicin only, 11% were resistant to isoniazid only, 58% were resistant to both isoniazid and rifampicin (MDR), 7% of the MDR isolates showed additional resistance to flouroquinolones (pre-XDR) while 5% of the MDR isolates were resistant to flouroquinolones and SLIDS (XDR). This study did not find any pre-XDR isolates with SLID resistance. The treatment was successful in 75% of the pre-XDR-TB cases, whereas all XDR-TB cases had unsuccessful treatment outcomes. All isoniazid monoresistant cases had unsuccessful treatment outcomes; 55% of the rifampicin monoresistant cases had unsuccessful treatment outcomes; among the MDR-TB cases, 73% had successful treatment outcomes (Figure 1). No statistically significant association was found between the second line drug resistance or the treatment outcome with HIV status, age, sex and *M.tb* family (Table 2). Table 3 shows characteristics and treatment outcomes of pre-XDR and XDR-TB cases in the study. The treatment outcomes for the rest of the cases are shown in supplementary Table S1. When evaluating the MTBDR*sl* results, it was found that the most common fluoroquinolone-resistance conferring mutation detected was *gyrA* A90V (found in 7% of the cases). The mutation *gyrA* G88A/G88C was only detected in one isolate. Among the pre-XDR-TB and XDR-TB cases, the second line injectable drug resistance was caused by the mutation *rrs* A1401G. Of the 7 pre-XDR and XDR-TB patients, the HIV status was not known for two patients, while the other five patients were HIV positive. Some patients with known HIV status had the same hybridization pattern, drug resistance profile, *M.tb* lineage and spoligo family as patients with unknown HIV status (Table 3).

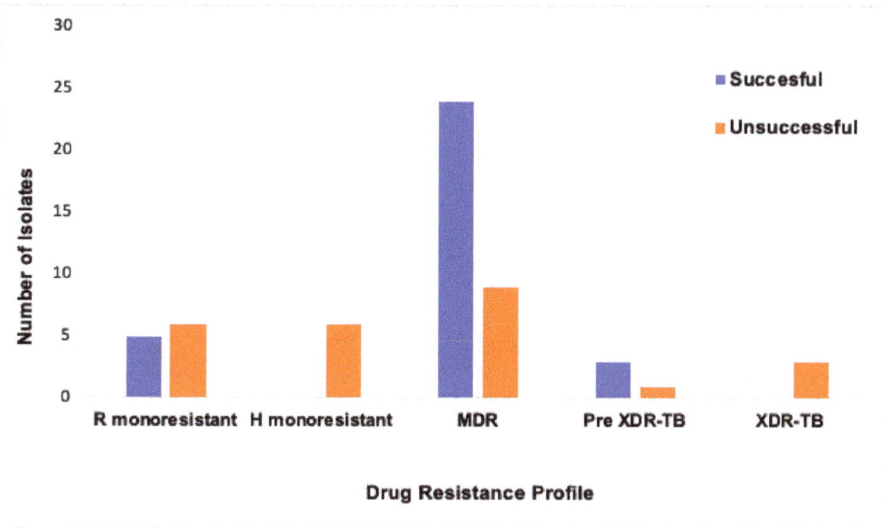

Figure 1. Treatment outcomes of patients with different drug resistance profiles.

Table 2. Factors associated with second line drug resistance.

	MDR N = 50	2nd Line Drug Resistance* N = 7	*p*-value
Sex	n (%)	n (%)	0.253
Male	27 (54)	2 (29)	
Female	23 (46)	5 (71)	
	Age in years		0.833
<20 years	8 (16)	1 (14)	
20–39 years	23 (47)	5 (71)	
40–59 years	15 (31)	1 (14)	
>60 years	3 (6)	0 (0)	
	HIV status		0.226
Negative	15 (30)	0 (0)	
Positive	26 (52)	5 (71)	
Unknown	9 (18)	2 (29)	
	Smear results		0.630
Negative	10 (20)	2 (29)	
Positive	40 (80)	5 (71)	
	Region		0.866
Central	20 (40)	4 (57)	
South West	1 (2)	0 (0)	
North West	24 (48)	0 (0)	
Southern	5 (10)	3 (43)	
	Lineage		0.066
Lineage 1	4 (8)	3 (43)	
Lineage 2	11 (22)	0 (0)	
Lineage 4	34 (68)	4 (57)	
Unknown	1 (2)	0 (0)	

*Second-line drug resistance includes pre-XDR and XDR-TB patients.

R- Rifampicin; H- isoniazid; MDR-TB- Multi-drug resistant TB; Pre-XDR-TB- Pre-extensively drug resistant tuberculosis; XDR-TB-extensively drug resistant tuberculosis. "Successful treatment"; includes patients who completed treatment and those who were cured; "Unsuccessful treatment"; includes patients who failed treatment, patients who are deceased, loss to follow up, defaulted, not evaluated and not initiated into treatment.

Table 3. Characteristics of XDR-TB and pre-XDR-TB cases detected among the drug resistant cases.

Case	Age	Sex	Region	HIV Status	FLDs Drug Resistance Pattern	Hybridization Pattern (s)	Codon Mutations	SLDs Drug Resistance Pattern	*M.tb* Lineage, Spoligo Family	Treatment Outcome
1	29	F	Central	Positive	H; R; S; E	*gyrA* ΔWT3 + *rrs* ΔWT1 + *rrs* MUT1	Undefined mutation, A1401G	OFL; LFX; KAN; AM; CAP	L4, X3	Failed
2	28	F	North West	Unknown	H; R; S; E	*gyrA* ΔWT3 + *rrs* ΔWT1 + *rrs* MUT1	Undefined mutation, A1401G	OFL; LFX; KAN; AM; CAP	L4, X3	Failed
3	32	M	North West	Positive	H; R; E	*gyrA* ΔWT2 + *gyrA MUT1*+ *rrs* ΔWT1 + *rrs* MUT1	A90V, A1401G	OFL; LFX; KAN; AM; CAP	L4, LAM4	Deceased
4	37	F	Central	Positive	H; R; E	*gyrA* ΔWT3 + *gyr*A MUT1	A90V	OFL; LFX	L1, EAI1_SOM	Completed
5	44	M	Central	Unknown	H; R; S; E	*gyrA* ΔWT2 + *gyrA* MUT1	A90V	OFL; LFX	L1, EAI1_SOM	*Not evaluated
6	34	F	Central	Positive	H; R; S; E	*gyrA* ΔWT2 + *gyrA* MUT1	A90V	OFL; LFX	L1, EAI1_SOM	Completed
7	16	F	South	Positive	H; R; S; E	*gyrA* ΔWT1	G88A/G88C	OFL; LFX	L4, LAM3	Cured

H: Isoniazid; R: Rifampicin; S: Streptomycin; E: Ethambutol; FLDs: first-line drugs; SLDs: second-line drugs; OFL: Ofloxacin; LFX: Levofloxacin; KAN: Kanamycin; AMK: Amikacin; CAP: Capreomycin: L4: Lineage 4; L1: Lineage 1. *Patient was not initiated on treatment. WT- wild type; ΔWT1, ΔWT2, ΔWT3- no hybridization at respective WT probe; MUT1-mutation.

4. Discussion

In this retrospective analysis of drug resistant isolates from Botswana, it is shown that there is a high proportion of rifampicin (R) and isoniazid (H) monoresistance posing an increased risk of the development of MDR-TB [21,22]. It was observed that 7% (4/57) of the isolates were pre-XDR-TB. These isolates had resistance to fluoroquinolone only while 5% (3/57) were XDR-TB. The majority, 88% (50/57) of the isolates did not have any resistance to second-line drugs. It is interesting to see that in this sample, set SLID resistance occurred after fluoroquinolone resistance and not the other way around. Fluoroquinolones are used to treat other bacterial infections other than TB which could play a role in the increasing levels of resistance to this class of drugs in *M.tb* (in both pre-XDR and XDR cases) [5]. The introduction of Pretomanid, recently approved by the Food and Drug Administration (FDA) for the treatment of R resistant TB, has greatly shortened treatment and has been seen to improve treatment success [23,24]. However, Pretomanid is not yet available in Botswana. The presence of both pre-XDR-TB and XDR-TB cases are indicators that there are gaps in the control of TB. The genotyping methods used in this study are not sufficiently discriminatory to investigate transmission, and whole genome sequencing has not been done on these strains. However, the possibility exists that patient to patient transmission exists as there were 2 patients who were infected with a strain of the same spoligotype and second-line drug resistance pattern. Further analysis would however be needed to investigate TB transmission in this population.

The GenoType MTBDR*sl* assay can detect mutations in the quinolone resistance determining region (QRDR) of the genes *gyrA* and *gyrB*. The most common *gyrA* mutation detected by the MTBDR*sl* among the pre-XDR and XDR cases was A90V (57%). This mutation confers resistance to levofloxacin and is associated with low level resistance to moxifloxacin [8]. Some resistance mutations are characterized by the absence of hybridization at the respective wild type probes [25]. The absence of the wild type bands in the line probe assay can be used to infer that there could be resistance to flouroquinolones but it does not allow the determination of the genotypic changes and the resulting phenotypic resistance to specific drugs. The targeted sequencing is therefore required to identify the specific drug resistance mutations. For example, in our cohort, there were two isolates which had an undefined mutation shown by the absence of both the wild type (*gyrA* WT3) and the mutation band in the *gyrA* gene. In this case, resistance to fluoroquinolones, particularly levofloxacin, can only be inferred since the specific mutation is not known [25]. This information can nevertheless help select a treatment regimen that could be more beneficial to the patients (e.g., excluding fluoroquinolones). The future implementation of new drugs, such as Bedaquiline and Pretomanid, could change the genetic drug resistance patterns observed in this study. The pre-XDR and XDR-TB patients with drug resistance patterns described in this study could still be successfully treated with these drugs, thereby reducing the spread of these *M.tb* strains. The genotypic drug resistance patterns therefore need to be closely monitored to be able to adapt treatment guidelines if required.

This study found that all the XDR-TB patients had unsuccessful treatment outcomes. These strains being XDR probably resulted in (almost) none of the prescribed drugs being efficient in killing the bacteria. The pre-XDR and XDR-TB patients in this study were managed with regimens which contained levofloxacin. Previous studies have shown that in cases of levofloxacin resistance, moxifloxacin may be the preferred drug of choice since *gyrA* A90V mutation has a smaller effect on moxifloxacin activity [8,26]. Therefore, these patients could have benefited from a regimen containing Moxifloxacin if the specific resistance markers had been determined timely. Among the pre-XDR isolates and MDR-TB isolates, 75% and 73% of the patients had successful treatment outcomes respectively, however all isoniazid monoresistant and 55% of the rifampicin monoresistant patients had unsuccessful treatment outcomes. Previous studies have shown that isoniazid monoresistance is associated with poor treatment outcomes [3,27].

There were no mutations detected in the *gyrB* gene in any of the isolates in this study. The mutations in the *gyrB* gene are usually associated with low level resistance to fluoroquinolones and are not as common as those in the *gyrA* gene [2]. The *rrs* MUT1 A1401G mutation which leads to a high level second-line injectable drug resistance was detected among 5% of the MDR-TB isolates. This mutation causes high level resistance to KAN and cross resistance to AM and CAP [6]. The presence of these mutations shows that there is a need to routinely test for second-line drug resistance among MDR-TB cases in Botswana. In this study, gene mutations that are associated with low level drug resistance induced by mutations in the promoter area of the *eis* gene were not detected. There was no association between the drug resistance profile and HIV status. The data on HIV viral load and CD4 cell counts were not available for this study and their association with drug resistance could therefore not be analyzed. However, previous studies have shown that there is an association between drug resistance and HIV status [28–30]. Haar et al. and Fenner et al. have shown that patients with high viral loads are more likely to have multi-drug resistant TB than those who are virally suppressed [31,32].

Even though this study is informative and provides data on the genetic mutations that are associated with second-line drug resistance in Botswana, it had some limitations. The small sample size may not reflect the true burden of second-line drug resistance in the entire country and there is limited statistical power to detect other drug resistance mutations within the population. Due to the small sample size, there is insufficient statistical power to fully address the association with various risk factors and treatment outcomes. This was a retrospective study therefore, there may be other unknown confounding factors. One of the limitations of the line probe assays (LPAs) is that there may be a false detection of resistance due to some synonymous mutations. Some studies have reported synonymous mutations which can result in false-positive results (false detection of resistance). However, in such instances, appropriate confirmatory testing should be done promptly [33]. The lack of hybridization of the wild type probes is not a reliable indication of phenotypic resistance, hence these kinds of hybridization patterns need to be verified with phenotypic DST [7]. There were some mutations that were undefined in our study therefore, other mechanisms of fluoroquinolone resistance need to be investigated further using techniques, such as whole genome sequencing or targeted gene sequencing. The identification of gyrase mutations can aid in predicting fluoroquinolone resistance as well as estimating the levels of resistance to various flouroquinolones. This may assist clinicians to determine the most effective dose of fluoroquinolones [34]. There is also a need to carry out this study in a larger population in order to determine the association of several risk factors with treatment outcomes as well as to determine the prevalence of second-line drug resistance in Botswana.

5. Conclusions

Our study shows that there is second-line drug resistance and the majority of cases had *gyrA* A90V, *rrs* A1401G mutation. Our results show that monitoring and further investigations with more discriminatory methods are required to determine whether these strains are transmitted or if second-line drug resistance is acquired during treatment.

Supplementary Materials: The following are available online at http://www.mdpi.com/2076-0817/8/4/208/s1, Table S1. Clinical characteristics and treatment outcomes of the drug resistant isolates in the study.

Author Contributions: Conceptualization: S.G., methodology: E.M.S., R.M.W., T.M., S.G., S.D.L., B.D.; formal analysis: T.M., P.M., S.M., E.M.S., R.M.W., S.D.L.; software: S.G.; validation: E.M.S., S.D.L., B.D., P.M.; visualization: E.M.S., S.D.L., B.D., T.M.; investigation: E.M.S., S.D.L., B.D., T.M.; resources: S.G., R.M.W.; data curation: S.G., B.K., G.R.-P., T.T.T., M.M.; writing original draft: T.M., S.D.L., writing review & editing: T.M., E.M.S., B.D., R.M.W., S.D.L., S.G., S.M., I.K., T.I., P.M., L.M., M.M., G.R.-P., T.T., B.K.; supervision: I.K., S.G.; project administration: S.G., T.M., I.K., R.M.W.; funding acquisition: S.G. All authors read and approved the final manuscript.

Funding: This work was supported through the Sub-Saharan African Network for TB/HIV Research Excellence (SANTHE), a DELTAS Africa Initiative [grant # DEL-15-006]. The DELTAS Africa Initiative is an independent funding scheme of the African Academy of Sciences (AAS)'s Alliance for Accelerating Excellence in Science in Africa (AESA) and supported by the New Partnership for Africa's Development Planning and Coordinating Agency (NEPAD Agency) with funding from the Wellcome Trust [grant # 107752/Z/15/Z] and the UK government. The views expressed in this publication are those of the author(s) and not necessarily those AAS, NEPAD Agency, Wellcome Trust or the UK government.

Acknowledgments: We would like to thank the National Tuberculosis Reference Laboratory and Ministry of Health and Wellness Botswana for providing clinical *M.tb* isolates for this study.

Conflicts of Interest: The authors declare no conflicts of interest.

References

1. WHO. Global Tuberculosis Report. Available online: https://www.who.int/tb/publications/global_report/en/ (accessed on 20 July 2019).
2. Oudghiri, A.; Karimi, H.; Chetioui, F.; Zakham, F.; Bourkadi, J.E.; Elmessaoudi, M.D.; Laglaoui, A.; Chaoui, I.; El Mzibri, M. Molecular characterization of mutations associated with resistance to second-line tuberculosis drug among multidrug-resistant tuberculosis patients from high prevalence tuberculosis city in Morocco. *BMC Infect. Dis.* **2018**, *18*, 98. [CrossRef] [PubMed]
3. Villegas, L.; Otero, L.; Sterling, T.R.; Huaman, M.A.; Van der Stuyft, P.; Gotuzzo, E.; Seas, C. Prevalence, risk factors, and treatment outcomes of isoniazid-and rifampicin-mono-resistant pulmonary tuberculosis in lima, peru. *PLoS ONE* **2016**, *11*, e0152933. [CrossRef] [PubMed]
4. O'Donnell, M. Isoniazid monoresistance: A precursor to multidrug-resistant tuberculosis? *Ann. Am. Thorac. Soc.* **2018**, *15*, 306–307. [CrossRef] [PubMed]
5. Nguyen, H.B.; Nguyen, N.V.; Tran, H.T.; Nguyen, H.V.; Bui, Q.T. Prevalence of resistance to second-line tuberculosis drug among multidrug-resistant tuberculosis patients in Viet Nam, 2011. *Western Pac. Surveill. Response J.* **2016**, *7*, 35–40. [CrossRef] [PubMed]
6. Hu, Y.; Hoffner, S.; Wu, L.; Zhao, Q.; Jiang, W.; Xu, B. Prevalence and genetic characterization of second-line drug-resistant and extensively drug-resistant Mycobacterium tuberculosis in Rural China. *Antimicrob. Agents Chemother.* **2013**, *57*, 3857–3863. [CrossRef]
7. Gardee, Y.; Dreyer, A.W.; Koornhof, H.J.; Omar, S.V.; da Silva, P.; Bhyat, Z.; Ismail, N.A. Evaluation of the GenoType MTBDRsl version 2.0 assay for second-line drug resistance detection of mycobacterium tuberculosis isolates in South Africa. *J. Clin. Microbiol.* **2017**, *55*, 791–800. [CrossRef]
8. Farhat, M.R.; Jacobson, K.R.; Franke, M.F.; Kaur, D.; Sloutsky, A.; Mitnick, C.D.; Murray, M. Gyrase mutations are associated with variable levels of fluoroquinolone resistance in mycobacterium tuberculosis. *J. Clin. Microbiol.* **2016**, *54*, 727–733. [CrossRef]
9. Hooper, D.C.; Jacoby, G.A. Mechanisms of drug resistance: Quinolone resistance. *Ann. N.Y. Acad. Sci.* **2015**, *1354*, 12–31. [CrossRef]
10. Jabeen, K.; Shakoor, S.; Malik, F.; Hasan, R. Fluoroquinolone resistance in Mycobacterium tuberculosis isolates from Pakistan 2010–2014: Implications for disease control. *Int. J. Mycobacteriol.* **2015**, *4*, 47–48. [CrossRef]
11. WHO. *The Use of Molecular Line Probe Assays for the Detection of Resistance to Second-Line Anti-Tuberculosis Drugs: Policy Guidance*; WHO: Geneva, Switzerland, 2016; Available online: https://www.who.int/tb/publications/lpa-mdr-diagnostics/en/ (accessed on 23 July 2019).
12. HainLifescience. *Instructions for Use for Hain GenoType MTBDRsl, Version 2.0*; HainLifescience: Nehren, Germany; Available online: https://www.hain-lifescience.de/en/instructions-for-use.html (accessed on 20 July 2019).
13. Tagliani, E.; Cabibbe, A.M.; Miotto, P.; Borroni, E.; Toro, J.C.; Mansjo, M.; Hoffner, S.; Hillemann, D.; Zalutskaya, A.; Skrahina, A.; et al. Diagnostic performance of the new version (v2.0) of genotype mtbdrsl assay for detection of resistance to fluoroquinolones and second-line injectable drugs: A multicenter study. *J. Clin. Microbiol.* **2015**, *53*, 2961–2969. [CrossRef]
14. Ismail, N.; Ismail, F.; Omar, S.V.; Blows, L.; Gardee, Y.; Koornhof, H.; Onyebujoh, P.C. Drug resistant tuberculosis in Africa: Current status, gaps and opportunities. *Afr. J. Lab. Med.* **2018**, *7*, 781. [CrossRef] [PubMed]

15. Nanzaluka, F.H.; Chibuye, S.; Kasapo, C.C.; Langa, N.; Nyimbili, S.; Moonga, G.; Kapata, N.; Kumar, R.; Chongwe, G. Factors associated with unfavourable tuberculosis treatment outcomes in Lusaka, Zambia, 2015: A secondary analysis of routine surveillance data. *Pan. Afr. Med. J.* **2019**, *32*, 159. [CrossRef]
16. Tanue, E.A.; Nsagha, D.S.; Njamen, T.N.; Assob, N.J.C. Tuberculosis treatment outcome and its associated factors among people living with HIV and AIDS in Fako Division of Cameroon. *PLoS ONE* **2019**, *14*, e0218800. [CrossRef] [PubMed]
17. Mogashoa, T.; Melamu, P.; Ley, S.D.; Streicher, E.M.; Iketleng, T.; Kelentse, N.; Mupfumi, L.; Mokomane, M.; Kgwaadira, B.; Novitsky, V.; et al. Genetic diversity of Mycobacterium tuberculosis strains circulating in Botswana. *PLoS ONE* **2019**, *14*, e0216306. [CrossRef]
18. National Tuberculosis Programme Manual 2011. Available online: https://www.who.int/hiv/pub/guidelines/botswana_tb.pdf (accessed on 5 January 2019).
19. HainLifescience. *Instructions for Use for Hain GenoLyse Kit for Isolation of Genomic Bacterial DNA*; HainLifescience: Nehren, Germany; Available online: https://www.hain-lifescience.de/en/instructions-for-use.html (accessed on 20 July 2019).
20. Kamerbeek, J.; Schouls, L.; Fau-Kolk, A.; Kolk, A.; Fau-van Agterveld, M.; van Agterveld, M.; Fau-van Soolingen, D.; van Soolingen, D.; Fau-Kuijper, S.; Kuijper, S.; et al. Simultaneous detection and strain differentiation of Mycobacterium tuberculosis for diagnosis and epidemiology. *J. Clin. Microbiol.* **1997**, *35*, 907–914. [PubMed]
21. Cox, H.; Kebede, Y.; Allamuratova, S.; Ismailov, G.; Davletmuratova, Z.; Byrnes, G.; Stone, C.; Niemann, S.; Rusch-Gerdes, S.; Blok, L.; et al. Tuberculosis recurrence and mortality after successful treatment: Impact of drug resistance. *PLoS Med.* **2006**, *3*, e384. [CrossRef]
22. Munang, M.L.; Kariuki, M.; Dedicoat, M. Isoniazid-resistant tuberculosis in Birmingham, United Kingdom, 1999–2010. *QJM* **2015**, *108*, 19–25. [CrossRef] [PubMed]
23. Kendall, E.A.; Malhotra, S.; Cook-Scalise, S.; Denkinger, C.M.; Dowdy, D.W. Estimating the impact of a novel drug regimen for treatment of tuberculosis: A modeling analysis of projected patient outcomes and epidemiological considerations. *BMC Infect. Dis.* **2019**, *19*, 794. [CrossRef]
24. Saravu, K.; Pai, M. Drug-resistant tuberculosis: Progress towards shorter and safer regimens. *Lung India* **2019**, *36*, 373–375. [CrossRef]
25. GLI. Line Probe Assays for Drug Resistant Tuberculosis Detection. Interpretation and Reporting Guide for Laboratory Staff and Clinicians. Available online: http://www.stoptb.org/wg/gli/assets/documents/LPA_test_web_ready.pdf (accessed on 20 July 2019).
26. Maitre, T.; Petitjean, G.; Chauffour, A.; Bernard, C.; El Helali, N.; Jarlier, V.; Reibel, F.; Chavanet, P.; Aubry, A.; Veziris, N. Are moxifloxacin and levofloxacin equally effective to treat XDR tuberculosis? *J. Antimicrob. Chemother.* **2017**, *72*, 2326–2333. [CrossRef]
27. Chien, J.Y.; Chen, Y.T.; Wu, S.G.; Lee, J.J.; Wang, J.Y.; Yu, C.J. Treatment outcome of patients with isoniazid mono-resistant tuberculosis. *Clin. Microbiol. Infect.* **2015**, *21*, 59–68. [CrossRef] [PubMed]
28. Yimer, S.A.; Agonafir, M.; Derese, Y.; Sani, Y.; Bjune, G.A.; Holm-Hansen, C. Primary drug resistance to anti-tuberculosis drugs in major towns of Amhara region, Ethiopia. *APMIS* **2012**, *120*, 503–509. [CrossRef] [PubMed]
29. Asmamaw, D.; Seyoum, B.; Makonnen, E.; Atsebeha, H.; Woldemeskel, D.; Yamuah, L.; Addus, H.; Aseffa, A. Primary drug resistance in newly diagnosed smear positive tuberculosis patients in Addis Ababa, Ethiopia. *Ethiop. Med. J.* **2008**, *46*, 367–374. [PubMed]
30. Vidyaraj, C.K.; Chitra, A.; Smita, S.; Muthuraj, M.; Govindarajan, S.; Usharani, B.; Anbazhagi, S. Prevalence of rifampicin-resistant Mycobacterium tuberculosis among human-immunodeficiency-virus-seropositive patients and their treatment outcomes. *J. Epidemiol. Glob. Health* **2017**, *7*, 289–294. [CrossRef] [PubMed]
31. Haar, C.H.; Cobelens, F.G.; Kalisvaart, N.A.; van der Have, J.J.; van Gerven, P.J.; van Soolingen, D. Tuberculosis drug resistance and HIV infection, the Netherlands. *Emerg. Infect. Dis.* **2007**, *13*, 776–778. [CrossRef]
32. Fenner, L.; Atkinson, A.; Boulle, A.; Fox, M.P.; Prozesky, H.; Zurcher, K.; Ballif, M.; Furrer, H.; Zwahlen, M.; Davies, M.A.; et al. HIV viral load as an independent risk factor for tuberculosis in South Africa: Collaborative analysis of cohort studies. *J. Int. AIDS Soc.* **2017**, *20*, 21327. [CrossRef]

33. Ajileye, A.; Alvarez, N.; Merker, M.; Walker, T.M.; Akter, S.; Brown, K.; Moradigaravand, D.; Schon, T.; Andres, S.; Schleusener, V.; et al. Some synonymous and nonsynonymous gyra mutations in mycobacterium tuberculosis lead to systematic false-positive fluoroquinolone resistance results with the hain genotype mtbdrsl assays. *Antimicrob. Agents Chemother.* **2017**, *61*. [CrossRef]
34. Disratthakit, A.; Prammananan, T.; Tribuddharat, C.; Thaipisuttikul, I.; Doi, N.; Leechawengwongs, M.; Chaiprasert, A. Role of gyrB mutations in pre-extensively and extensively drug-resistant tuberculosis in thai clinical isolates. *Antimicrob. Agents Chemother.* **2016**, *60*, 5189–5197. [CrossRef]

Article

Genotyping, Assessment of Virulence and Antibacterial Resistance of the Rostov Strain of *Mycobacterium tuberculosis* Attributed to the Central Asia Outbreak Clade

Mikhail V. Fursov [1,*], Egor A. Shitikov [2,*], Julia A. Bespyatykh [2], Alexander G. Bogun [1], Angelina A. Kislichkina [1], Tatiana I. Kombarova [1], Tatiana I. Rudnitskaya [1], Natalia S. Grishenko [1], Elena A. Ganina [1], Lubov V. Domotenko [1], Nadezhda K. Fursova [1], Vasiliy D. Potapov [1,†] and Ivan A. Dyatlov [1,†]

1. State Research Center for Applied Microbiology and Biotechnology, Obolensk 142279, Russia; bogun@obolensk.org (A.G.B.); angelinakislichkina@yandex.ru (A.A.K.); kombarova.tatyana@yandex.ru (T.I.K.); rudnitskaya.tat@yandex.ru (T.I.R.); natalya.grishenko60@mail.ru (N.S.G.); ganin43@yandex.ru (E.A.G.); domotenko@obolensk.org (L.V.D.); n-fursova@yandex.ru (N.K.F.); potapovvd@mail.ru (V.D.P.); dyatlov@obolensk.org (I.A.D.)
2. Federal Research and Clinical Center of Physical-Chemical Medicine, Moscow 119435, Russia; juliabespyatykh@gmail.com

* Correspondence: mikhail.fursov88@gmail.com (M.V.F.); eshitikov@mail.ru (E.A.S.)

† Equally managed the research.

Received: 28 February 2020; Accepted: 29 April 2020; Published: 30 April 2020

Abstract: The Central Asia Outbreak (CAO) clade is a growing public health problem for Central Asian countries. Members of the clade belong to the narrow branch of the *Mycobacterium tuberculosis* Beijing genotype and are characterized by multidrug resistance and increased transmissibility. The Rostov strain of *M. tuberculosis* isolated in Russia and attributed to the CAO clade based on PCR-assay and whole genome sequencing and the laboratory strain H37Rv were selected to evaluate the virulence on C57Bl/6 mice models by intravenous injection. All mice infected with the Rostov strain succumbed to death within a 48-day period, while more than half of the mice infected by the H37Rv strain survived within a 90-day period. Mice weight analysis revealed irreversible and severe depletion of animals infected with the Rostov strain compared to H37Rv. The histological investigation of lung and liver tissues of mice on the 30th day after injection of mycobacterial bacilli showed that the pattern of pathological changes generated by two strains were different. Moreover, bacterial load in the liver and lungs was higher for the Rostov strain infection. In conclusion, our data demonstrate that the drug-resistant Rostov strain exhibits a highly virulent phenotype which can be partly explained by the CAO-specific mutations.

Keywords: Beijing genotype; Central Asia Outbreak; murine infection model; Virulence; pre-XDR-TB

---

1. Introduction

Worldwide, tuberculosis (TB) is a major public health problem. Despite a slight decrease in TB incidence rates in recent years (1.6% per year in the period 2000–2018 and 2.0% between 2017 and 2018), the situation remains extremely tense. A total of 1.5 million people died from the disease and more than 9 million new cases were detected in 2018 worldwide [1]. Currently, seven lineages of *Mycobacterium tuberculosis* are described, which can cause disease and demonstrate specific phylogeographic patterns [2]. Of them, lineage 2 and lineage 4 are the most widely dispersed, affecting humans across the world. Lineage 2 (or East Asian lineage) is arguably the most widespread and the Beijing genotype family is its major component (13% of global *M. tuberculosis* population; predominant in East Asia and Northern Eurasia) [3,4].

Multiple clinical and epidemiological studies demonstrated a strict association of Beijing genotype members with a high level of drug resistance combined with a large number of compensatory mutations, as well as enhanced pathogenicity, which lead to increased transmissibility and rapid progression of infection [5–7]. However, virulence studies provide less conclusive results, showing a variety of phenotypes. The latter is confirmed in animal models [8,9] and in vitro models of macrophage infection [10,11]. This is due to the fact that hypervirulence is not a characteristic feature of the Beijing genotype, but is specific only for certain genetic sublineages, often associated with disease outbreaks in some regions [12,13]. One of the most detailed examples is the spread of the virulent *M. tuberculosis* Beijing B0/W148 cluster in the Russian territory [14].

Phylogenetically, members of lineage 2 may be assigned to at least two large branches, termed ancient and modern sublineages, and 11 populations belong to these branches [15]. Modern Beijing sublineage strains are prevalent worldwide, leading to speculation that this sublineage has hypervirulent features [16]. In the present study, we aimed to investigate a strain belonging to a more homogenous group within modern Beijing called Central Asia Outbreak (CAO) clade. This clade is a part of the Central Asian population. The latter was initially designated as CC1 or Central Asian [6] and then as East Europe 1 [17] and Central Asian and Russian [18]. It largely correlates with the 94-32 cluster and M2 subtype according to multilocus variable-number tandem-repeat analysis (MLVA) and mycobacterial interspersed repetitive unit (MIRU) typing, respectively [19,20]. The members of the population are characterized by a high level of drug resistance and comprise about one-fourth of the pathogen population in Uzbekistan, Tajikistan, Kyrgyzstan, and Kazakhstan [21–23]. Additionally, these strains distributed in Russia and other former Soviet Republics [7]. It should be noted, that this population is heterogeneous and includes at least two large clades: CladeA and the previously mentioned CAO. CladeA strains are prevalent in Russia. In turn, CAO isolates are not often identified in Russia and are usually associated with the spread of resistant TB forms in the former Soviet Central Asia [7,21].

In the current study, we examine genetic and phenotypic characteristics of the Rostov strain of *M. tuberculosis* belonging to the CAO clade of the Beijing genotype. This strain was attributed to the pre-extensively drug-resistant (XDR) tuberculosis group. The growth rate and virulence for mice of the Rostov strain were compared with the same characteristics of the *M. tuberculosis* H37Rv strain.

2. Results

2.1. Genetic and Phenotypic Characteristics of the Strain

The Rostov strain of *Mycobacterium tuberculosis* was isolated in the South Federal District of Russia in 2013 from a patient with pulmonary tuberculosis.

The Beijing genotype (SIT1) was confirmed by spoligotyping. The assignment of the strain to a CAO clade of the Central Asian and Russian Beijing population (Beijing 94-32 cluster) was revealed by PCR assays as in [18,23], respectively. The next generation sequencing analysis on Ion Torrent additionally confirmed that the strain belongs to the CAO clade of the Beijing genotype and carries all previously described CAO-specific single nucleotide polymorphisms (SNPs) [21]. 24-locus MIRU-variable-number tandem-repeat (VNTR) typing scheme revealed 223325153533424682254423 profile which is designated 9358-25 in the international MIRU-VNTRplus database. This profile is closest to the 94-32 and has two different loci, as shown in Figure S1.

According to the drug susceptibility testing and genome analysis, the strain belonged to pre-XDR tuberculosis and was resistant to streptomycin, isoniazid, rifampicin, ethambutol, kanamycin, amikacin, and capreomycin, as shown in Table 1. Additionally, a putative compensatory mutation in the *rpoC* gene (g764363a; G332S) was revealed.

Table 1. MIC distribution of the Rostov strain of *Mycobacterium tuberculosis* for antibiotics and drug resistance markers.

Antibiotic	MIC, mg/L	Interpretation	Drug Resistance Marker
Isoniazid (INH)	>1	R	KatG (S315T)
Rifampicin (RIF)	>40	R	RpoB (S450L)
Streptomycin (STR)	>10	R	RpsL (K43R)
Ethambutol (EMB)	>5	R	EmbB (M306V)
Amikacin (AMK)	>30	R	*rrs* (a1401g)
Kanamycin (KAN)	>30	R	*rrs* (a1401g)
Capreomycin (CAP)	>30	R	*rrs* (a1401g)
Ofloxacin (OFX)	≤3	S	-

Note: MIC—minimal inhibitory concentration; R—resistance; S—sensitivity.

To compare the growth rate between the Rostov and H37Rv strains we determined a growth index and C_{max} value, as shown in Figure 1. Growth index reflects that the Rostov strain grew faster than the H37Rv strain throughout the experimental period ($p < 0.05$), as shown in Figure 1A. The Rostov strain showed a higher C_{max} than the H37Rv strain ($p < 0.05$), as shown in Figure 1B. C_{max} for the Rostov strain was reached on the 25th day, whilst C_{max} for the H37Rv strain was reached on the 30th day.

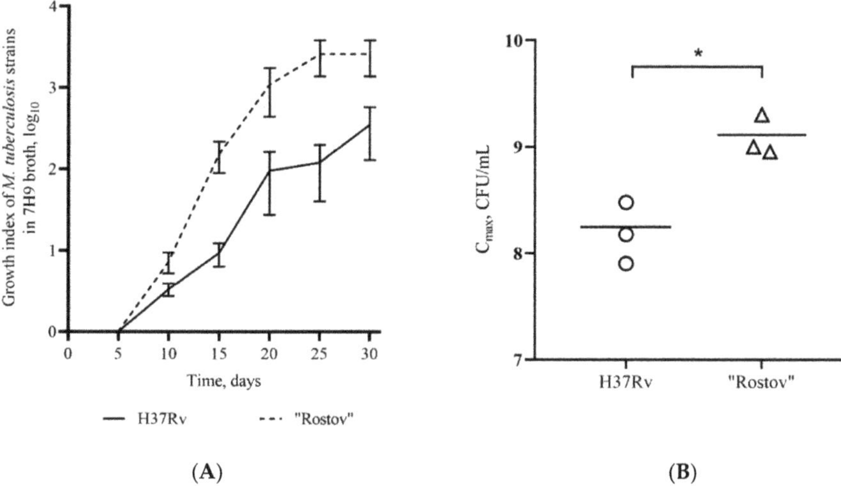

Figure 1. Growth of H37Rv and Rostov strains of *M. tuberculosis* in 7H9 broth. **(A)**—Growth index (calculated by the colony-forming unit (CFU) at each time point divided by the CFU at initial time point); **(B)**—Comparison of C_{max} (a maximum point on the growth curve); *—values of $p < 0.05$.

2.2. Mice Survival Rate and Bodyweight Dynamic

The model of *M. tuberculosis* infection of C57BL/6 mice was used for the comparison of the virulence of the Rostov clinical strain and the reference virulent strain H37Rv. Animals were intravenously injected with 5×10^6 CFU/mice of each strain (nine mice per strain, $n = 18$). Additionally, as a negative control, a group of uninfected animals ($n = 9$) was used. The patterns of animal survival were observed from the first to 90 days post-infection (p.i.). In each group of mice, weight control was performed. Figure 2 shows that mice infected with the Rostov and H37Rv strains started to die after 18 and 36 days of infection, respectively. All mice of the group infected with the Rostov strain succumbed to death within a 47-day period, while ~56% of mice infected by the H37Rv strain survived within a 90-day p.i. period. Mice weight analysis showed irreversible and severe depletion of animals infected with the clinical Rostov strain compared to animals infected with the laboratory H37Rv strain, as shown in Figure 3.

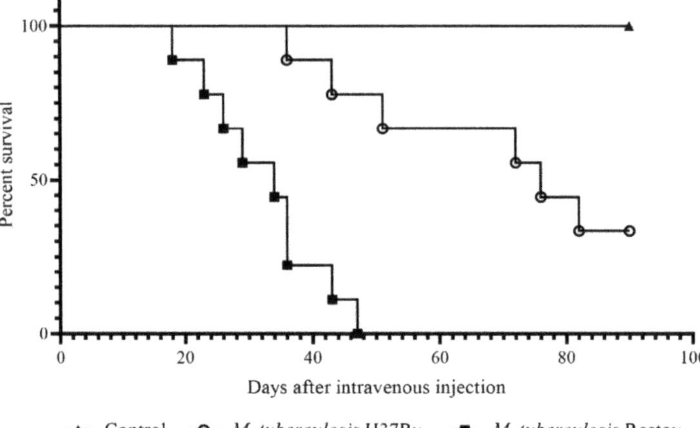

Figure 2. Comparison of the survival curves of C57BL/6 mice infected by *M. tuberculosis* strains. Data were analyzed by the Gehan–Breslow–Wilcoxon test. The value of $p < 0.05$ was taken as statistically significant.

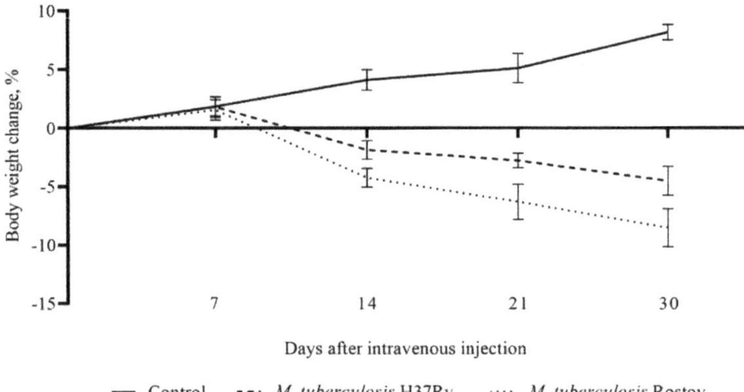

Figure 3. Comparison of the weight changes curves of C57BL/6 mice infected by *M. tuberculosis* strains. The value of $p < 0.05$ was taken as statistically significant.

2.3. Investigation of Tuberculosis Process on the 30th Day of Infection

To further define, the virulence of the studied strains, we investigated the tuberculosis process in the C57BL/6 mice models on the 30th day after pathogen injection, when all mice infected by the H37Rv strain were alive, and about 50% of mice infected by the Rostov strain were dead. The pathological processes provided by two *M. tuberculosis* strains were very different. Animal appearance after infection by the H37Rv strain was characterized by mild depletion and smooth fur, but after infection by the Rostov strain—by extremely emaciated and "ruffled" fur. The differences in survival times were associated with differences in the macroscopic appearance of lungs and liver harvested on the 30th day of infection, when more than 50% of mice infected with the Rostov strain were dead. It was shown that the lungs of mice infected by the Rostov strain were different from those in the H37Rv-infected mice, which appeared in increased lungs volume, intensively hyperemic, and no visible nodules; in turn, the lungs of mice infected by the H37Rv strain were pale pink colored with pale mass inclusions. The similar picture was obtained in the liver: Rostov-infected mice livers were dark brown with multiple nodules and the fatty degeneration was visible, while the livers from H37Rv-infected mice were smooth, dark brown, and normal volume, as shown in Table 2.

Table 2. Comparative characterization of the mortality, animal appearance and morphological description of internal organs of C57Bl/6 mice infected by the H37Rv and Rostov strains of *M. tuberculosis*.

Strain	The Mortality Rate on 30th p.i. Day, %	Animal Appearance	Lungs	Liver
H37Rv	0	Mild depletion, smooth fur	Pale pink colored with pale mass inclusions	Smooth, intense brown, normal volume
Rostov	~50%	Extreme emaciated, hunched posture, "ruffled" fur, reduced movement	Intensively hyperemic, no visible nodules	Dark brown with multiple nodules, fatty degeneration

The histological investigation of the C57Bl/6 mice infected intravenously by the H37Rv strain of *M. tuberculosis* at a dose of 5×10^6 CFU/animal showed a typical picture for TB mice models in the lungs: the small granulomas composed of numerous macrophages with abundant cytoplasm form; there are some lymphocytes between macrophages; dense perivascular lymphocytic infiltrates form in the lungs in addition to granulomas, as shown in Figure 4A. A single infiltrate consisting of few lymphocytes was found in histological sections of the liver, as shown in Figure 4C.

In contrast, when the C57Bl/6 mice were infected by the Rostov strain of *M. tuberculosis*, the pattern of pathological changes was different. Diffuse thickening of the alveolar septum due to an increased number of macrophages occurred in certain parts of the lungs. Lymphocytic infiltrates were not observed, as shown in Figure 4B. Microscopy of histological sections of the liver showed the presence of nodules consisting of focal cell accumulations in the parenchyma. Cellular infiltrates are composed of few typical macrophages and a large number of polymorphonuclear leukocytes, that indicate intensive pathogen multiplication and the increased development of a pathological inflammatory process, as shown in Figure 4D.

Bacterial load in the lungs and liver of mice infected with the Rostov and H37Rv strains was measured on day 30 p.i. According to the data presented in Figure 5, the clinical Rostov strain more actively proliferate in the parenchymatous organs of experimental animals than the H37Rv strains. The overall bacterial load in the lungs was higher than in the liver for both strains.

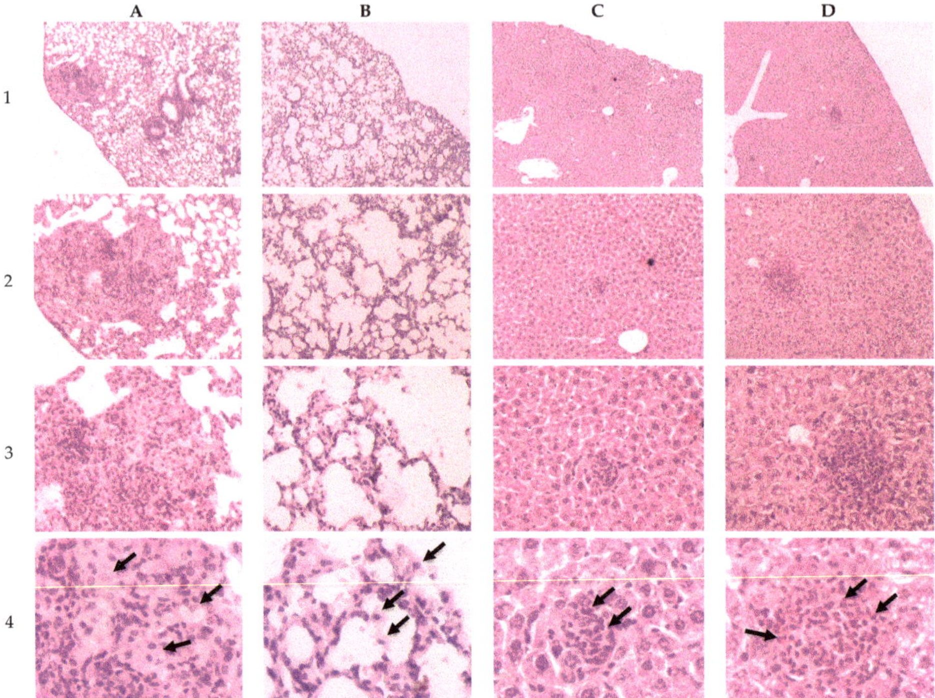

Figure 4. Histology of lungs and livers of C57Bl/6 mice on the 30th day after intravenous inoculation by the *M. tuberculosis* strains H37Rv (**A**—lungs, **C**—liver) and Rostov (**B**—lungs, **D**—liver). 1, 2, 3 and 4—×4, ×10, ×20 and ×40 magnification, respectively. The arrow indicates the specific mice cells (**A**4, **B**4—macrophages; **C**4—lymphocytes; **D**4—polymorphonuclear leukocytes).

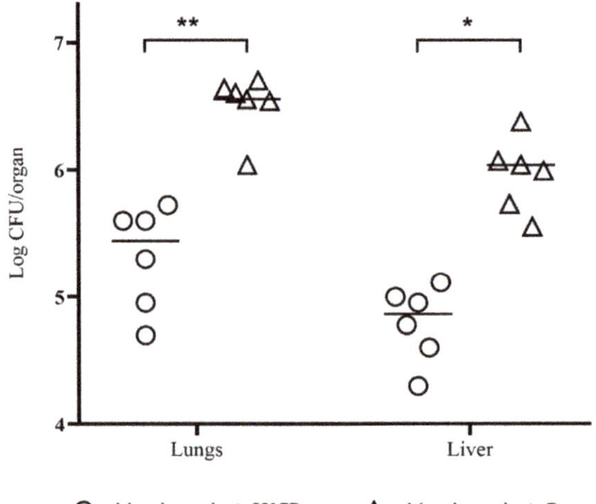

Figure 5. *M. tuberculosis* cells loads of the C57Bl/6 mice lungs and liver on the 30th day after the inoculation of bacteria; *—values of $p \leq 0.05$; **—values of $p \leq 0.01$.

3. Discussion

In order to better understand the virulence properties of CAO strains, we focused on the clinical Rostov strain belonging to the clade. The strain was resistant to seven antituberculosis drugs and contained well-known resistance-associated mutations, as shown in Table 1. Resistance to OFX was not detected for the strain that correlates with a study of Merker et al. [21] in which the frequency of resistance to fluoroquinolones was low among CAO isolates. Besides the drug resistance-associated mutations, the strain carried a compensatory mutation in the *rpoC* gene (g764363a; G332S), which was previously described [7,24]. We suggest that this mutation could affect the fitness and lead to an increased growth rate of the strain compared to the reference H37Rv strain shown in Figure 1, in contrast with data published for 3 strains of lineage 2, that had decreased growth rate compared to the same reference strain [25].

According to 24-locus MIRU-VNTR typing *M. tuberculosis*, Rostov belonged to the 9358-25 cluster and differed from the 94-32 cluster by two loci, as shown in Figure S1. Although this cluster was not described earlier, the phylogenetic analysis using the MIRU-VNTR-plus database revealed a clusterization with the 94-32 type and according to MIRU typing, it belongs to the M2 cluster that is specific to Central Asia population [18].

Analysis of cluster-specific SNPs revealed one significant point mutation (a2321369g; N105D) in the *Rv2063a* (MazF7) gene related to virulence, detoxification and adaptation category according to Mycobrowser database (https://mycobrowser.epfl.ch/) and Forrellad et al. [26]. The second specific SNP was identified in the *fadE29* gene resulting in an amino acid substitution Ile288Val. Such substitution did not provide the significant changes in protein structure, accordingly to BLOSUM62 Matrix [27] (Table S1). It was reported previously that the MazEF toxin–antitoxin system is very important for stress adaptation, drug tolerance, and virulence of *M. tuberculosis*, and required for persistence in vitro. The deletion of MazF reduced the pathogen virulence for guinea pigs and decreased the bacterial load in organs [28]. All other polymorphisms presented in the Table S1 are not specific for the CAO clade, but their role is likely to be important for successful spread of the Beijing genotype in the world.

Survival studies showed that mice infected with Rostov strain succumbed to death within 18–47 days p.i., whereas a large proportion of mice infected with H37Rv maintained viability up to 90 days p.i., as shown in Figure 2. Similar mortality rates were detected for Beijing *M. tuberculosis* strains; conversely, the strains belonging to other *M. tuberculosis* families—Canetti, Haarlem and Somali clades—displayed intermediate or low virulence according to Lopez et al. [9].

Analysis of the specific pulmonary lesions in mice with experimental tuberculosis on day 30 showed that both strains had characteristic pathogenic properties, i.e., were able to cause the tuberculosis process, but patterns of pathological changes in lungs and livers were different for two strains, as shown in Figure 4. Our results are in agreement with Ribeiro et al. [29], according to which the H37Rv strain had the least virulent properties with respect to the Beijing genotype. The obtained data indicate that infection of mice with the clinical Rostov strain of *M. tuberculosis* leads to changes in the lungs. These changes consist of a small increase in the number of macrophages in some interalveolar septums. At the same time, macrophages have a relatively narrow cytoplasm, in contrast to wide-plasma macrophages that infiltrate lung tissue when mice are infected with the H37Rv strain of *M. tuberculosis*. The absence of pulmonary infiltrates in mice infected with the Rostov strain may indicate that this strain did not activate the host defense mechanisms, compared with the response to the infection caused by the H37Rv strain, as shown in Figure 4.

In conclusion, our study showed that pre-XDR Rostov strain belonging to the CAO clade of *M. tuberculosis* Beijing genotype is characterized by high virulence for C57Bl/6 mice when compared with the laboratory H37Rv strain. We propose that characteristic alterations of the CAO clade favor the selection of highly virulent bacteria.

4. Materials and Methods

4.1. M. tuberculosis Strains

The Rostov strain of *M. tuberculosis* was initially isolated from a clinical sample of sputum collected from a 35-year-old man hospitalized in the South Federal District of Russia in 2013 and deposed into the State Collection of Pathogenic Microorganisms "SCPM-Obolensk" (ID B-7601). The virulent laboratory H37Rv strain of *M. tuberculosis* was obtained from the "SCPM-Obolensk" collection (ID B-4825).

Frozen stocks of bacterial cells (1×10^5 CFU) were inoculated into 30 mL Middlebrook 7H9 broth with OADC supplement (BD, Franklin Lakes, NJ, USA), and 0.05% Tween 80, in three biological replicates, incubated at 37 °C under static conditions (i.e., without agitation) in flask cell culture 250 mL (Greiner AG, Kremsmünster, Austria) for 30 days to estimate the growth rate. Every five days the aliquots of 0.1 mL were taken for CFU enumeration by plating the serial 10-fold dilutions in triplicates onto Middlebrook 7H11 agar (BD, Franklin Lakes, NJ, USA) enriched with OADC. Bacterial colonies were counted on the plates after incubation for three weeks at 37 °C. To compare the growth rate amongst strains, we determined a growth index, calculated from the $\log_{10}$ of the number of CFU at each time point divided by the $\log_{10}$ of the number of CFU at the initial time point. Additionally, we used C_{max} to compare the growth of the strains. This index means the peak point on the bacterial growth curve [25].

4.2. Antibacterial Susceptibility

M. tuberculosis drug susceptibility testing (DST) of the Rostov strain to isoniazid 1.0 mg/L (INH), rifampin 40.0 mg/L (RIF), streptomycin 10.0 mg/L (STR), ethambutol 5.0 mg/L (EMB), amikacin 30.0 mg/L (AMK), kanamycin 30.0 mg/L (KAN), capreomycin 30.0 mg/L (CAP), and ofloxacin 3.0 mg/L (OFX) was carried out using the method of absolute concentrations on solid Lowenstein–Jensen medium [30]. In addition, DST was performed by the BACTEC MGIT 960 system (BD, Sparks, MD, USA) according to the manufacturer's instructions.

4.3. Genomic Analysis

Genomic DNA was isolated from the Rostov strain of *M. tuberculosis* using a standard extraction method [31].

Spoligotyping and 24 MIRU-VNTR typing were performed as described in references [32,33], respectively. Verification of the presence of SNP in the *sigA* gene and CAO-specific IS*6110* insertion in the *Rv1359-Rv1360* intergenic region was performed by PCR as described previously [18,34].

Whole genome sequencing was performed on Ion Torrent PGM (Life Technologies, Camarillo, CA, USA) with Ion 318 chip and Ion PGM™ Sequencing 200 Kit v2 (Life Technologies, Camarillo, CA, USA). Raw sequence data were submitted to the NCBI under the project PRJNA269675. The genome was assembled using Newbler GS de novo assembler 2.5 (Roche, Branford, CT, USA) with standard parameters for Ion technology. SNPs were detected with Snippy v.4.3.6 (https://github.com/tseemann/snippy) pipeline with a minimum coverage depth of 10 and an alternate fraction of 0.9. A comprehensive list of drug-resistance mutations to first- and second-line drugs was used to determine genetic resistance of the strain [35]. Functional categories and virulence factors were defined according to Mycobrowser (https://mycobrowser.epfl.ch/) and [27].

4.4. Bioethical Requirements

All animal experiments were carried out in full accordance with the European Convention for the Protection of Vertebrate Animals, used for experimental and other scientific purposes (Directive 2010/63/EU of the European Parliament and of the Council of 22 September 2010 on the protection of animals used for scientific purposes), and the requirements of Sanitary Regulations 1.3.2322-08 "Safety of work with microorganisms of the III-IV pathogenicity groups and pathogens of parasitic diseases", and Veterinary Protocol No. VP-2016/8 were approved by bioethics of the State Research Center for Applied Microbiology and Biotechnology.

4.5. Mice Infection

Specified pathogen-free female C57BL/6 mice (n = 51) were obtained from Shemyakin and Ovchinnikov Institute of Bioorganic Chemistry RAS (Moscow, Russia). All mice were used at 7–8 weeks of age and 20–22 g in weight. Randomization was used to allocate three experimental groups: control, *M. tuberculosis* H37Rv, and *M. tuberculosis* Rostov.

M. tuberculosis strains were grown to mid-logarithmic phase (OD_{600} = 1.0), cells were collected by centrifugation, and washed with PBS containing 0.05% Tween-80. Mice were intravenously injected into the lateral tail vein with 5×10^6 CFU/mice (in 0.1 mL of 0.9% NaCl) of the H37Rv strain and of the Rostov strain. All animals were weighed each day after infection. Animals were observed for 90 days; the physical appearance and behavior of animals were estimated; the daily animal weight loss and mortality were calculated.

On the 30th day after *M. tuberculosis* injection, six mice were euthanized by CO_2 gas in each experimental group. Lungs and livers tissues were examined for mycobacterial load and pathology. The *M. tuberculosis* bacillary burden in lungs and livers was counted by homogenates plating onto Middlebrook 7H11 agar. Some samples of lungs and livers were fixed in 10% formalin (BioChem-NN, Nizhny Novgorod, Russia), graded concentrations of ethanol and butanol were used for dehydration, embedded in paraffin, and serial sections (5 μm width) were prepared with the Ultracut microtome (Reichert-Jung, Bensheim, Germany). Sections were deparaffinated and stained with hematoxylin and eosin. All slides were examined with a Nikon Eclipse 80i microscope and a Nikon DS-U2digital camera (Nikon, Tokyo, Japan).

4.6. Statistical Methods

Analysis of data was conducted using GraphPad Prism version 8.0.1 for Windows (GraphPad Software, La Jolla, CA, USA, www.graphpad.com). Statistical analysis between groups was performed using the analysis of variance (ANOVA) test. Survival data were analyzed using the Gehan–Breslow–Wilcoxon test. The growth index and C_{max} were compared amongst the H37Rv and Rostov strains using the unpaired t-test at each time point. A value of $p < 0.05$ was considered significant.

Supplementary Materials: The following are available online at http://www.mdpi.com/2076-0817/9/5/335/s1, Figure S1. MIRU-VNTRplus cluster analysis of *M. tuberculosis* Rostov strain, Table S1. Sublineage-specific nsSNPs in the virulence genes.

Author Contributions: Conceptualization, M.V.F. and E.A.S.; methodology, V.D.P., and N.K.F.; software, A.G.B., A.A.K. and J.A.B.; validation, L.V.D., T.I.K. and T.I.R.; formal analysis, A.A.K. and N.K.F.; investigation, N.S.G., T.I.K., T.I.R. and E.A.G.; resources, I.A.D. and V.D.P.; data curation, A.G.B.; writing—original draft preparation, M.V.F., E.A.S. and J.A.B.; writing—review and editing N.K.F.; visualization, E.A.G., and M.V.F.; supervision, V.D.P. and I.A.D.; project administration, V.D.P.; funding acquisition, I.A.D. All authors have read and agreed to the published version of the manuscript.

Funding: This work was supported by a grant from the Ministry of Science and Higher Education of the Russian Federation No. 075-15-2019-1671 (agreement dated 31 October 2019).

Acknowledgments: The authors are thankful to Khramov, M.V. and Sectoral Scientific Program of the Russian Federal Service for Surveillance on Consumer Rights Protection and Human Wellbeing.

Conflicts of Interest: The authors declare no competing financial interest.

References

1. *WHO Global Tuberculosis Report 2019*; World Health Organization: Geneva, Switzerland, 2019.
2. Comas, I.; Coscolla, M.; Luo, T.; Borrell, S.; Holt, K.E.; Kato-Maeda, M.; Parkhill, J.; Malla, B.; Berg, S.; Thwaites, G.; et al. Out-of-Africa migration and Neolithic coexpansion of Mycobacterium tuberculosis with modern humans. *Nat. Genet.* **2013**, *45*, 1176–1182. [CrossRef] [PubMed]

3. Wiens, K.E.; Woyczynski, L.P.; Ledesma, J.R.; Ross, J.M.; Zenteno-Cuevas, R.; Goodridge, A.; Ullah, I.; Mathema, B.; Djoba Siawaya, J.F.; Biehl, M.H.; et al. Global variation in bacterial strains that cause tuberculosis disease: A systematic review and meta-analysis. *BMC Med.* **2018**, *16*, 196. [CrossRef] [PubMed]
4. Brudey, K.; Driscoll, J.R.; Rigouts, L.; Prodinger, W.M.; Gori, A.; Al-Hajoj, S.A.; Allix, C.; Aristimuño, L.; Arora, J.; Baumanis, V.; et al. Mycobacterium tuberculosis complex genetic diversity: Mining the fourth international spoligotyping database (SpolDB4) for classification, population genetics and epidemiology. *BMC Microbiol.* **2006**, *6*, 23. [CrossRef] [PubMed]
5. Parwati, I.; van Crevel, R.; van Soolingen, D. Possible underlying mechanisms for successful emergence of the Mycobacterium tuberculosis Beijing genotype strains. *Lancet Infect. Dis.* **2010**, *10*, 103–111. [CrossRef]
6. Merker, M.; Blin, C.; Mona, S.; Duforet-Frebourg, N.; Lecher, S.; Willery, E.; Blum, M.G.B.; Rüsch-Gerdes, S.; Mokrousov, I.; Aleksic, E.; et al. Evolutionary history and global spread of the Mycobacterium tuberculosis Beijing lineage. *Nat. Genet.* **2015**, *47*, 242–249. [CrossRef]
7. Casali, N.; Nikolayevskyy, V.; Balabanova, Y.; Harris, S.R.; Ignatyeva, O.; Kontsevaya, I.; Corander, J.; Bryant, J.; Parkhill, J.; Nejentsev, S.; et al. Evolution and transmission of drug-resistant tuberculosis in a Russian population. *Nat. Genet.* **2014**, *46*, 279–286. [CrossRef]
8. Aguilar, D.; Hanekom, M.; Mata, D.; Gey Van Pittius, N.C.; Van Helden, P.D.; Warren, R.M.; Hernandez-Pando, R. Mycobacterium tuberculosis strains with the Beijing genotype demonstrate variability in virulence associated with transmission. *Tuberculosis* **2010**, *90*, 319–325. [CrossRef]
9. López, B.; Aguilar, D.; Orozco, H.; Burger, M.; Espitia, C.; Ritacco, V.; Barrera, L.; Kremer, K.; Hernandez-Pando, R.; Huygen, K.; et al. A marked difference in pathogenesis and immune response induced by different Mycobacterium tuberculosis genotypes. *Clin. Exp. Immunol.* **2003**, *133*, 30–37. [CrossRef]
10. Li, Q.; Whalen, C.C.; Albert, J.M.; Larkin, R.; Zukowski, L.; Cave, M.D.; Silver, R.F. Differences in rate and variability of intracellular growth of a panel of Mycobacterium tuberculosis clinical isolates within a human monocyte model. *Infect. Immun.* **2002**, *70*, 6489–6493. [CrossRef]
11. Theus, S.A.; Cave, M.D.; Eisenach, K.D. Intracellular macrophage growth rates and cytokine profiles of Mycobacterium tuberculosis strains with different transmission dynamics. *J. Infect. Dis.* **2005**, *191*, 453–460. [CrossRef]
12. Bifani, P.J.; Mathema, B.; Kurepina, N.E.; Kreiswirth, B.N. Global dissemination of the Mycobacterium tuberculosis W-Beijing family strains. *Trends Microbiol.* **2002**, *10*, 45–52. [CrossRef]
13. Hanekom, M.; Van Der Spuy, G.D.; Streicher, E.; Ndabambi, S.L.; McEvoy, C.R.E.; Kidd, M.; Beyers, N.; Victor, T.C.; Van Helden, P.D.; Warren, R.M. A recently evolved sublineage of the Mycobacterium tuberculosis Beijing strain family is associated with an increased ability to spread and cause disease. *J. Clin. Microbiol.* **2007**, *45*, 1483–1490. [CrossRef] [PubMed]
14. Mokrousov, I. Insights into the origin, emergence, and current spread of a successful Russian clone of Mycobacterium tuberculosis. *Clin. Microbiol. Rev.* **2013**, *26*, 342–360. [CrossRef]
15. Shitikov, E.; Kolchenko, S.; Mokrousov, I.; Bespyatykh, J.; Ischenko, D.; Ilina, E.; Govorun, V. Evolutionary pathway analysis and unified classification of East Asian lineage of Mycobacterium tuberculosis. *Sci. Rep.* **2017**, *7*, 1–10. [CrossRef]
16. Liu, Q.; Luo, T.; Dong, X.; Sun, G.; Liu, Z.; Gan, M.; Wu, J.; Shen, X.; Gao, Q. Genetic features of Mycobacterium tuberculosis modern Beijing sublineage. *Emerg. Microbes Infect.* **2016**, *5*, e14. [CrossRef]
17. Luo, T.; Comas, I.; Luo, D.; Lu, B.; Wu, J.; Wei, L.; Yang, C.; Liu, Q.; Gan, M.; Sun, G.; et al. Southern East Asian origin and coexpansion of Mycobacterium tuberculosis Beijing family with Han Chinese. *Proc. Natl. Acad. Sci. USA* **2015**, *112*, 8136–8141. [CrossRef]
18. Mokrousov, I.; Chernyaeva, E.; Vyazovaya, A.; Skiba, Y.; Solovieva, N.; Valcheva, V.; Levina, K.; Malakhova, N.; Jiao, W.W.; Gomes, L.L.; et al. Rapid assay for detection of the epidemiologically important Central Asian/Russian strain of the Mycobacterium tuberculosis Beijing genotype. *J. Clin. Microbiol.* **2018**, *56*, e01551-17. [CrossRef]
19. Narvskaya, O.; Mokrousov, I.; Otten, T.; Vishnevskiy, B. Molecular markers: Application for studies of Mycobacterium tuberculosis population in Russia. In *Trends in DNA Fingerprinting Research*; Read, M.M., Ed.; Nova Science Publishers: New York, NY, USA, 2005.
20. Mokrousov, I.; Narvskaya, O.; Vyazovaya, A.; Millet, J.; Otten, T.; Vishnevsky, B.; Rastogi, N. Mycobacterium tuberculosis Beijing genotype in Russia: In search of informative variable-number tandem-repeat loci. *J. Clin. Microbiol.* **2008**, *46*, 3576–3584. [CrossRef]

21. Merker, M.; Barbier, M.; Cox, H.; Rasigade, J.P.; Feuerriegel, S.; Kohl, T.A.; Diel, R.; Borrell, S.; Gagneux, S.; Nikolayevskyy, V.; et al. Compensatory evolution drives multidrug-resistant tuberculosis in central Asia. *Elife* **2018**, *7*, e38200. [CrossRef]
22. Engström, A.; Antonenka, U.; Kadyrov, A.; Kalmambetova, G.; Kranzer, K.; Merker, M.; Kabirov, O.; Parpieva, N.; Rajabov, A.; Sahalchyk, E.; et al. Population structure of drug-resistant Mycobacterium tuberculosis in Central Asia. *BMC Infect. Dis.* **2019**, *19*, 908. [CrossRef]
23. Skiba, Y.; Mokrousov, I.; Ismagulova, G.; Maltseva, E.; Yurkevich, N.; Bismilda, V.; Chingissova, L.; Abildaev, T.; Aitkhozhina, N. Molecular snapshot of Mycobacterium tuberculosis population in Kazakhstan: A country-wide study. *Tuberculosis* **2015**, *95*, 538–546. [CrossRef]
24. Song, T.; Park, Y.; Shamputa, I.C.; Seo, S.; Lee, S.Y.; Jeon, H.S.; Choi, H.; Lee, M.; Glynne, R.J.; Barnes, S.W.; et al. Fitness costs of rifampicin resistance in Mycobacterium tuberculosis are amplified under conditions of nutrient starvation and compensated by mutation in the β' subunit of RNA polymerase. *Mol. Microbiol.* **2014**, *91*, 1106–1119. [CrossRef]
25. Sarkar, R.; Lenders, L.; Wilkinson, K.A.; Wilkinson, R.J.; Nicol, M.P. Modern lineages of Mycobacterium tuberculosis exhibit lineage-specific patterns of growth and cytokine induction in human monocyte-derived macrophages. *PLoS ONE* **2012**, *7*, e43170. [CrossRef]
26. Forrellad, M.A.; Klepp, L.I.; Gioffré, A.; Sabio García, J.; Morbidoni, H.R.; de la Paz Santangelo, M.; Cataldi, A.A.; Bigi, F. Virulence factors of the Mycobacterium tuberculosis complex. *Virulence* **2013**, *4*, 3–66. [CrossRef]
27. Henikoff, S.; Henikoff, J.G. Amino acid substitution matrices from protein blocks (amino add sequence/alignment algorithms/data base srching). *Proc. Natl. Acad. Sci. USA* **1992**, *89*, 10915–10919. [CrossRef]
28. Tiwari, P.; Arora, G.; Singh, M.; Kidwai, S.; Narayan, O.P.; Singh, R. MazF ribonucleases promote Mycobacterium tuberculosis drug tolerance and virulence in guinea pigs. *Nat. Commun.* **2015**, *6*, 6059. [CrossRef]
29. Ribeiro, S.C.M.; Gomes, L.L.; Amaral, E.P.; Andrade, M.R.M.; Almeida, F.M.; Rezende, A.L.; Lanes, V.R.; Carvalho, E.C.Q.; Suffys, P.N.; Mokrousov, I.; et al. Mycobacterium tuberculosis strains of the modern sublineage of the Beijing family are more likely to display increased virulence than strains of the ancient sublineage. *J. Clin. Microbiol.* **2014**, *52*, 2615–2624. [CrossRef]
30. Canetti, G.; Froman, S.; Grosset, J.; Hauduroy, P.; Langerova, M.; Mahler, H.T.; Meissner, G.; Mitchison, D.A.; Sula, L. Mycobacteria: Laboratory methods for testing drug sensitivity and resistance. *Bull. World Health Organ.* **1963**, *29*, 565–578.
31. Van Embden, J.D.A.; Cave, M.D.; Crawford, J.T.; Dale, J.W.; Eisenach, K.D.; Gicquel, B.; Hermans, P.; Martin, C.; McAdam, R.; Shinnick, T.M.; et al. Strain identification of Mycobacterium tuberculosis by DNA fingerprinting: Recommendations for a standardized methodology. *J. Clin. Microbiol.* **1993**, *31*, 406–409. [CrossRef]
32. Kamerbeek, J.; Schouls, L.; Kolk, A.; Van Agterveld, M.; Van Soolingen, D.; Kuijper, S.; Bunschoten, A.; Molhuizen, H.; Shaw, R.; Goyal, M.; et al. Simultaneous detection and strain differentiation of Mycobacterium tuberculosis for diagnosis and epidemiology. *J. Clin. Microbiol.* **1997**, *35*, 907–914. [CrossRef]
33. Supply, P.; Allix, C.; Lesjean, S.; Cardoso-Oelemann, M.; Rüsch-Gerdes, S.; Willery, E.; Savine, E.; De Haas, P.; Van Deutekom, H.; Roring, S.; et al. Proposal for standardization of optimized mycobacterial interspersed repetitive unit-variable-number tandem repeat typing of Mycobacterium tuberculosis. *J. Clin. Microbiol.* **2006**, *44*, 4498–4510. [CrossRef] [PubMed]
34. Shitikov, E.; Vyazovaya, A.; Malakhova, M.; Guliaev, A.; Bespyatykh, J.; Proshina, E.; Pasechnik, O.; Mokrousov, I. Simple assay for detection of the central Asia outbreak clade of the mycobacterium tuberculosis Beijing genotype. *J. Clin. Microbiol.* **2019**, *57*, e00215-19. [CrossRef] [PubMed]
35. Schleusener, V.; Köser, C.U.; Beckert, P.; Niemann, S.; Feuerriegel, S. Mycobacterium tuberculosis resistance prediction and lineage classification from genome sequencing: Comparison of automated analysis tools. *Sci. Rep.* **2017**, *7*, 1–9. [CrossRef] [PubMed]

Article

Immune Phenotype and Functionality of *Mtb*-Specific T-Cells in HIV/TB Co-Infected Patients on Antiretroviral Treatment

Lucy Mupfumi [1,2,*], Cheleka A.M. Mpande [3], Tim Reid [3], Sikhulile Moyo [2,4], Sanghyuk S. Shin [5], Nicola Zetola [6,7], Tuelo Mogashoa [1], Rosemary M. Musonda [2,4], Ishmael Kasvosve [1], Thomas J. Scriba [3], Elisa Nemes [3,†] and Simani Gaseitsiwe [2,4,*,†]

1 Department of Medical Laboratory Sciences, Faculty of Health Sciences, University of Botswana, Gaborone, Botswana; tuelomogashoa@me.com (T.M.); kasvosvei@ub.ac.bw (I.K.)
2 Botswana Harvard AIDS Institute Partnership, Gaborone 320, Botswana; sikhulilemoyo@gmail.com (S.M.); rmusonda@gmail.com (R.M.M.)
3 South African Tuberculosis Vaccine Initiative, Institute of Infectious Disease and Molecular Medicine, Division of Immunology, Department of Pathology, University of Cape Town, Cape Town 7935, South Africa; MPNCHE001@myuct.ac.za (C.A.M.M.); tim.reid@uct.ac.za (T.R.); thomas.scriba@uct.ac.za (T.J.S.); elisa.nemes@uct.ac.za (E.N.)
4 Department of Immunology and Infectious Diseases, Harvard T.H. Chan School of Public Health, Boston, MA 02138, USA
5 Sue & Bill Gross School of Nursing, University of California Irvine, Irvine, CA 92697, USA; ssshin2@uci.edu
6 Infectious Diseases Division, University of Pennsylvania, Philadelphia, PA 19104, USA; nzetola@gmail.com
7 Botswana-Upenn Partnership, Gaborone, Botswana
* Correspondence: lmupfumi@gmail.com (L.M.); sgaseitsiwe@gmail.com (S.G.)
† Contributed equally.

Received: 21 December 2019; Accepted: 27 February 2020; Published: 2 March 2020

Abstract: The performance of host blood-based biomarkers for tuberculosis (TB) in HIV-infected patients on antiretroviral therapy (ART) has not been fully assessed. We evaluated the immune phenotype and functionality of antigen-specific T-cell responses in HIV positive (+) participants with TB (n = 12) compared to HIV negative (−) participants with either TB (n = 9) or latent TB infection (LTBI) (n = 9). We show that the cytokine profile of Mtb-specific CD4+ T-cells in participants with TB, regardless of HIV status, was predominantly single IFN-γ or dual IFN-γ/TNFα. Whilst ESAT-6/CFP-10 responding T-cells were predominantly of an effector memory (CD27−CD45RA−CCR7−) profile, HIV-specific T-cells were mainly of a central (CD27+CD45RA−CCR7+) and transitional memory (CD27+CD45RA+/−CCR7−) phenotype on both CD4+ and CD8+ T-cells. Using receiving operating characteristic (ROC) curve analysis, co-expression of CD38 and HLA-DR on ESAT-6/CFP-10 responding total cytokine-producing CD4+ T-cells had a high sensitivity for discriminating HIV+TB (100%, 95% CI 70–100) and HIV−TB (100%, 95% CI 70–100) from latent TB with high specificity (100%, 95% CI 68–100 for HIV−TB) at a cut-off value of 5% and 13%, respectively. TB treatment reduced the proportion of *Mtb*-specific total cytokine+CD38+HLA-DR+ CD4+ T-cells only in HIV−TB (p = 0.001). Our results suggest that co-expression of CD38 and HLA-DR on *Mtb*-specific CD4+ T-cells could serve as a TB diagnosis tool regardless of HIV status.

Keywords: immune activation; HLA-DR; CD38; treatment response

---

1. Introduction

Tuberculosis (TB) results in an estimated 1.5 million deaths each year, making it the number one cause of death by a single infectious organism [1]. The burden of TB is disproportionately higher in

Africa, where the co-infection rate with HIV is also the highest; more than 50% of the TB cases are HIV positive (+) [1]. The risk of active TB increases up to 30 times with HIV infection, with an annual risk of 15% compared to a 10% lifetime risk in HIV-uninfected participants [2]. Although antiretroviral therapy (ART) significantly reduces this risk [3], the incidence of TB while on ART remains higher than that in HIV-uninfected individuals [4,5]. HIV also specifically targets *Mycobacterium tuberculosis* (*Mtb*)-specific CD4 T-cells, resulting in impaired immune responses to TB in co-infected individuals [6,7].

Untreated HIV infection induces changes in the activation, memory, and functional profile of both CD4+ and CD8+ T-cells, with a decrease in naïve cells, and accumulation of highly differentiated cells [8]. It is hypothesized that this depletion of memory cells sustains the incidence of opportunistic infections in participants with advanced HIV. Uncontrolled replication of *Mtb* is associated with loss of CD27 expression [9], probably related to increased cellular homing to sites of disease. Recently, a ratio based on CD27 median fluorescent intensity (MFI) on CD4+ T-cells compared to that on IFN-γ+CD4+ T-cells has been shown to differentiate active disease from infection [10]. However, whether this approach has good diagnostic potential in HIV+TB remains to be determined.

Studies have also shown that activated effector *Mtb*-specific CD4+ T-cells characterized by the phenotype CD38+HLA-DR+ effectively distinguish latent TB infection (LTBI) from active TB cases [11–13] Furthermore, the frequency of *Mtb*-specific CD4+ T-cells expressing CD38 and HLA-DR declined rapidly within the first month of anti-tuberculosis treatment (ATT) [11,14]. CD38 expression on bulk CD8+ T-cells has long been recognized as a marker for disease progression in HIV-infected participants [15]. In addition, activated HLA-DR+ bulk and antigen-specific CD4+ T-cells have been shown to be a risk factor for TB progression in BCG-vaccinated infants [16] and to effectively discriminate latent from active TB in adults, respectively [13,14]. However, these studies were conducted predominantly in HIV− or HIV+ ART naïve individuals. It remains unclear what the impact of combined ART and ATT on the activation profile of *Mtb*-specific T-cells is.

ART seems to partially correct the HIV-induced T-cell defects and the activation profile of bulk T-cells from patients on ART is higher than that in HIV-uninfected individuals [17]. In a study of HIV-infected women who were followed up for a year post ART initiation, viral suppression and CD4+ T-cell increase did not lead to a concomitant decline in early (CD27+CD45RO+) and terminally differentiated (CD27−CD45RO+) memory CD4+ T-cells [17]. Since the memory phenotype determines the extent to which pathogen-specific responses are restored while on ART, persistent defects in the memory subsets of *Mtb*-specific CD4+ T-cells could account for the risk of active TB while on ART [18]. Therefore, it remains unclear how long-term ART affects the normalization of *Mtb*-specific T-cell memory phenotypes. Thus, in order to understand the dynamics of *Mtb*-specific T-cells during combined ART and ATT, we assessed the functional, activation, and differentiation profile of *Mtb*-specific T-cells in HIV/TB co-infected participants on ART compared to HIV− active TB. We also assessed whether the effects of ATT on immune responses were restricted to *Mtb*-specific T-cells by adding a Gag stimulation condition as a control for HIV-specific responses.

2. Results

2.1. Cohort Characteristics

We restricted analysis to participants who had samples at TB diagnosis and two months post ATT to allow the determination of changes in the activation and functional phenotype of *Mtb*-specific T-cells due to ATT. We enrolled 43 participants with pulmonary TB and excluded samples from 14 participants due to poor specimen viability (n = 6) and technical (instrument/processing) errors (n = 8, Figure 1). A total of 30 samples were analyzed: 12 HIV positive (HIV+TB), 9 HIV negative (HIV−TB) and 9 HIV-LTBI. HIV+TB participants had a median age of 39 years (Q1, Q3: 33, 48) and had been on ART for a median of 1.3 years (Q1, Q3 0.2, 6, Table 1). All participants reported a cough at baseline and had culture negative results after two months of ATT.

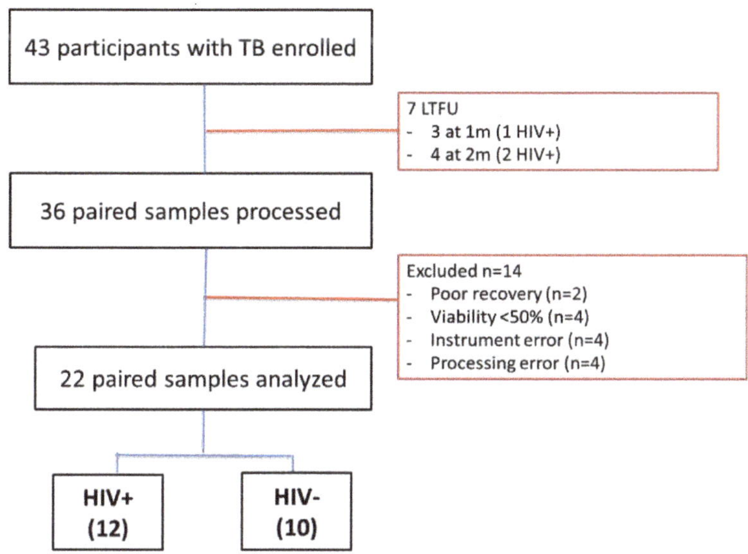

Figure 1. Study flow diagram. Samples were excluded for the following reasons: no follow-up sample available (n = 7), poor sample viability (n = 6), or technical errors (n = 8).

Table 1. Baseline characteristics of the study participants.

Characteristic	HIV+TB (n = 12)	HIV−TB (n = 9)	LTBI (n = 9)
Age, median years (Q1, Q3)	39 (33, 48)	28 (25, 40)	31 (28, 36)
Female (%)	7 (58)	2 (22)	4 (44)
Previous TB diagnosis	2 (17)	0	0
Cough duration (%)			N/A
<1 m	8 (67)	0	
1–2	3 (25)	4 (44)	
2–3	1 (8)	3 (33)	
>3 m	0	2 (22)	
CD4 T-cell count, median (Q1, Q3)	288 (172, 357)	N/A	N/A
CD4:CD8 ratio, median (Q1, Q3)	0.77 (0.4, 1.12)	N/A	N/A
Viral load, copies/mL [1] median (Q1, Q3)	317,212 (76, 327,725)	N/A	N/A
ART duration, median years (Q1, Q3)	1.3 (0.2, 6)	N/A	N/A

[1] Viral load results for three participants with detectable viral load at baseline; one had been on ART for 1.9 years and two were not on ART at the time, but were initiated on Dolutegravir (DTG)-based ART two weeks after ATT. By the two-month study visit, all patients had viral load results below 40 copies/mL.

2.2. Single IFN-γ and Dual IFN-γ/TNFα Response Characterizes Active Tuberculosis Regardless of HIV Status

We compared the magnitude of cytokine responses between TB and LTBI participants. Representative dot plots of the cytokine response to the four stimulation conditions are shown in the top panel of Figure 2A. At the time of TB diagnosis, IL-2−IFN-γ+TNFα+CD4+ T-cell responses to ESAT-6/CFP-10 stimulation were significantly higher in TB compared to LTBI but did not differ by HIV status (Figure 2B).

In a similar manner, single IFN-γ+ responses were higher for both HIV+TB (p = 0.01) and HIV−TB (p = 0.001) compared to LTBI (Figure 2). There was no difference in the frequencies of polyfunctional (p = 0.12) or single IFN-γ responses between HIV+TB and HIV−TB (p = 0.33). This was also reflected in the combinatorial polyfunctionality analysis of antigen-specific T-cells (COMPASS) heatmap (Figure 3A), which showed similar probabilities for the polyfunctional and dual expression of IFN-γ/TNFα for both TB groups. We compared the polyfunctionality score (PFS) across groups, which summarizes the functional profile of each participant by calculating the posterior probabilities of antigen-specific responses [19]. There was no difference in the polyfunctional scores (PFS) by either TB or HIV status (p = 0.07, Figure 3B). Interestingly, when we analyzed Gag stimulated CD8+ T-cells, no polyfunctional responses were observed, instead, responses were characterized by single IFN-γ expression (Supplementary Figure S2A). We also found an increase in the PFS of ESAT-6/CFP-10-stimulated CD4+ T-cells at 2 months post ATT only in the HIV+TB (p = 0.02 compared to baseline PFS, Figure 3C), but not in HIV−TB (p = 0.72, Figure 3D). We did not observe any other differences in the cytokine profile between the baseline and two-month timepoint (results not shown). These data suggest that the frequency of polyfunctional *Mtb*-specific T-cells may not immediately reflect the reduction in mycobacterial burden due to TB treatment.

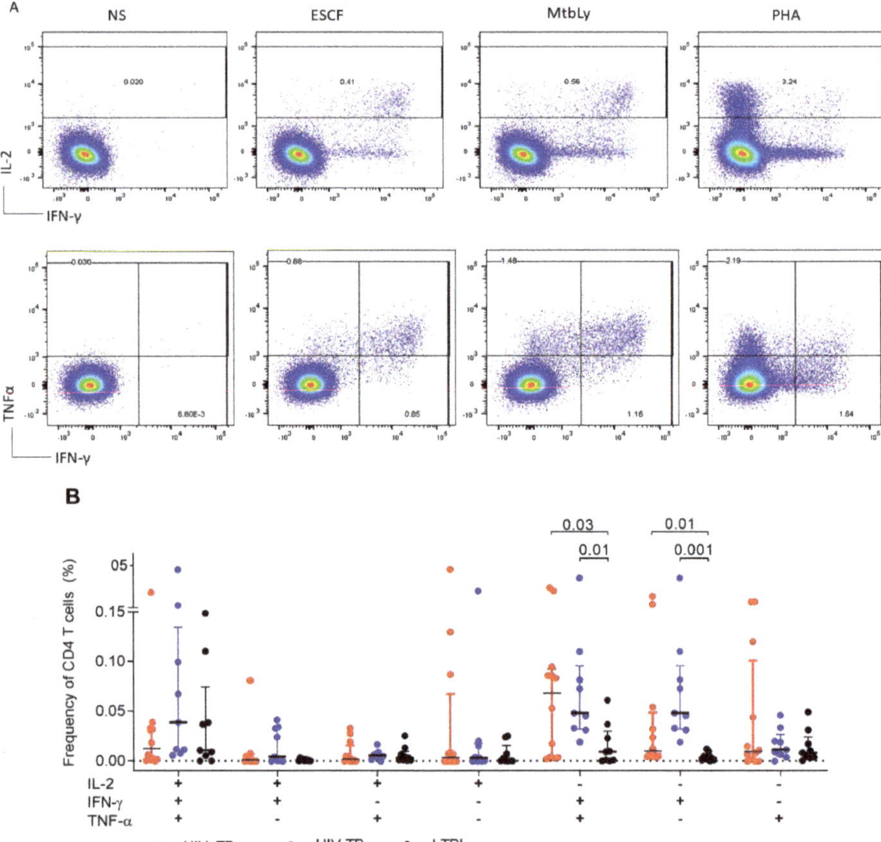

Figure 2. Comparison of the polyfunctional capacity of ESAT-6/CFP-10 CD4+ T-cells between HIV+TB, HIV−TB, and LTBI. (**A**) Representative plot of the expression of IL-2, *IFN-γ*, and TNFα in an HIV+TB participant. NS = unstimulated, ESCF = ESAT-6/CFP-10, MtbLy = Mtb-Lysate, PHA = Phytohemaglutinin. (**B**) Frequencies of CD4+ T-cells producing any of the possible combinations of IL-2, *IFN-γ*, and TNFα in response to ESAT-6/CFP-10 stimulation. Horizontal bars show the median and first and third quartile (Q1, Q3). Only statistically significant differences by the Kruskal–Wallis test with Dunn's correction for multiple comparisons are shown on the graph.

We also compared CD4 responses in the HIV+TB participants across stimulation conditions and observed that the relative proportions of polyfunctional and single TNFα response were highest on

Mtb-Lysate-stimulated cells compared to Gag ($p = 0.03$; Figure 4A). In addition, dual IFN-γ/TNFα was higher on Mtb-Lysate ($p = 0.002$) and ESAT-6/CFP-10 ($p = 0.05$) compared to Gag-stimulated CD4+ T-cells (Figure 4B). This was also reflected on the COMPASS heatmap that showed similar PFS with no distinct cytokine patterns separating the groups analyzed (Supplementary Figure S2B). We further analyzed the functional profile of *Mtb*-specific CD4+ T-cells by comparing the MFI of TNFα on total cytokine+ cells. While the TNFα MFI on ESAT-6/CFP-10 stimulated CD4+ T-cells did not differ by HIV or TB status ($p = 0.15$, results not shown), we observed a decline at two months of TB treatment in HIV−TB ($p = 0.02$) but not HIV+TB ($p = 0.27$) (Supplementary Figure S3).

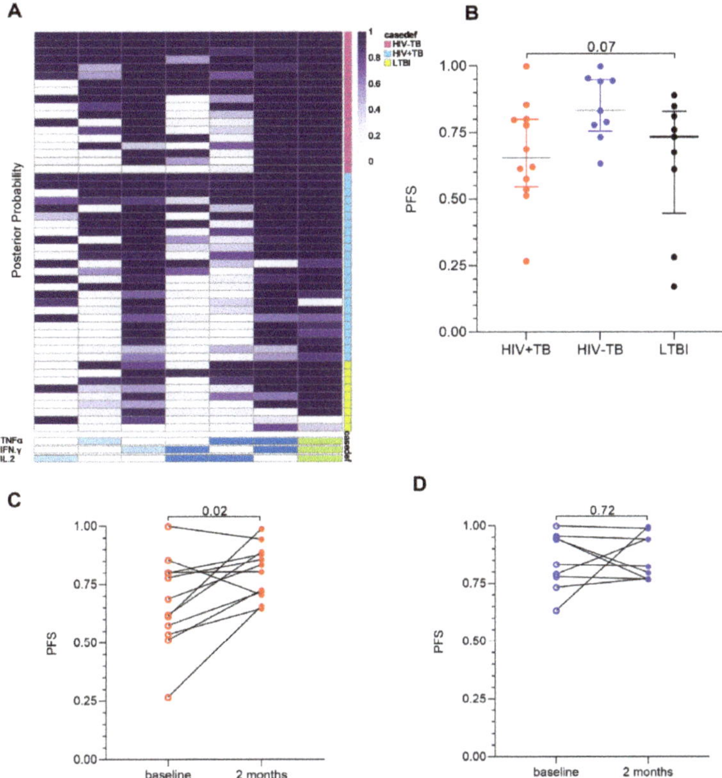

Figure 3. Polyfunctionality analysis of ESAT-6/CFP-10 specific CD4+ T-cell responses (**A**) Heatmap of combinatorial polyfunctionality analysis of antigen-specific T-cells (COMPASS) posterior probabilities of the distribution of responses in ESAT-6/CFP-10 stimulated CD4+ T-cells. Columns correspond to the different cell subsets modeled using the COMPASS package in R [19], color coded by the cytokines they express (white = "off", shaded = "on", grouped by color = "degree of polyfunctionality") ordered by degree of functionality from one function on the left to all three functions on the right (shaded from light blue to green on the bottom). Rows correspond to participants; one on each line, the color of each cell represents the probability (range 0–1) that the cell subset exhibits an antigen-specific response. Each cell of the heatmap shows the probability that a given cell-subset (column) has an antigen-specific response in the corresponding participant (the groups are coded on the right of the heatmap; pink = HIV+TB, blue = HIV−TB, and green = LTBI). Plots of polyfunctional scores (PFS) stratified by patient group (**B**) and timepoint for HIV+TB (**C**) and HIV−TB (**D**). Red dots represent HIV+TB, blue dots HIV−TB, and black dots LTBI. Differences between groups were compared using the Kruskal–Wallis test with Dunn's correction for multiple comparisons are shown on the graph. We assessed differences between timepoints using the Wilcoxon signed-rank test.

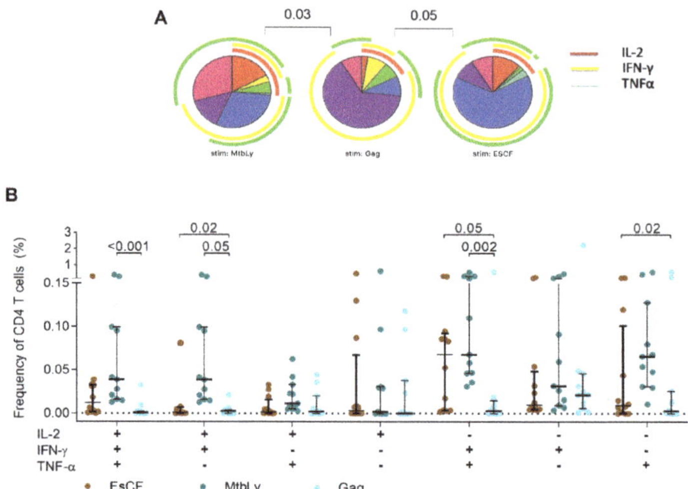

Figure 4. Functional analysis of HIV+TB. The functional profile of CD4+ T-cells specific for different antigens was assessed for CD4+ T-cells from HIV+TB patients at the time of TB diagnosis. (**A**) Each slice of the pie corresponds to a distinct combination of cytokines shown in (**B**) and defined by Boolean analysis in FlowJo. A key to the colors used for the arcs in the pie charts is shown on the top right. Pie charts were compared using the SPICE permutation test with $p < 0.05$ considered significant. (**B**) Frequency of cells producing any possible combination of IFN-γ, IL-2, or TNFα in response to the different stimulations. Horizontal bars represent median and interquartile range (IQR). Statistical comparisons were made using Kruskal–Wallis with Dunn's test for multiple comparisons. MtbLy = Mtb-Lysate; ESCF = ESAT-6/CFP-10.

2.3. Frequencies of CD38+HLA-DR+CD4+ T-cells Are Elevated in Active TB Participants Regardless of HIV Status and Decline after Two Months of TB Treatment

We hypothesized that the activation profile of *Mtb*-specific CD4+ T-cells might differ by HIV status. We evaluated the expression of CD38, HLA-DR, and Ki67 alone or in combination on ESAT-6/CFP-10, Mtb-Lysate, and Gag-stimulated total cytokine+CD4+ T-cells. Representative dot plots of the expression of CD38, HLA-DR, and Ki67 on total cytokine CD4+ T-cells are shown in Figure 5A. All individuals tested had a positive CD4+T-cell response to Mtb-Lysate while 9/12 (75%) HIV+TB and 8/9 (89%) LTBI individuals had responses to ESAT-6/CFP10. We observed on ESAT-6/CFP-10-stimulated CD4+ T-cells that the triple expression of CD38+HLA-DR+Ki67+ was associated with active TB only in HIV+TB participants when compared to LTBI ($p = 0.01$), although there was no difference between HIV+TB and HIV−TB ($p = 0.12$; Figure 5B). On Mtb-Lysate stimulated cells, differences in the expression of CD38+HLA-DR+KI67+ CD4+ T-cells were observed between HIV+TB and LTBI ($p = 0.002$) and HIV−TB and LTBI ($p = 0.04$, Figure 5C).

In addition, the predictive ability of the triple activation profile in discriminating active from latent TB was high only in HIV+TB (AUC = 0.86, 95% CI 0.68–1.00, $p = 0.01$, Figure 6A), corresponding to a sensitivity of 78% (95% CI 45–96) and a specificity of 75% (95% CI 41–96) at a cut-off of 1.8%. The sensitivity in HIV−TB was low at the same cut-off (33% vs. 78%) with comparable specificity (75%) and an AUC = 0.68 (95% CI 0.42–0.96, $p = 0.19$, Figure 6B). Sensitivity improved with Mtb-Lysate stimulation to 100% (95% CI 74–100) for HIV+TB with similar specificity (80%, 95% CI 38–99) at a cut-off value of 0.30 (results not shown). In contrast, CD38+HLA-DR+Ki67− expression on ESAT-6/CFP-10-stimulated CD4+ T-cells had a high sensitivity for both HIV+TB (100%, 95% CI 70–100) and HIV−TB (100%, 95% CI 70–100) and modest specificity (63%, 95% CI 31–86 for HIV+TB and 88%, 95% CI 53–99 for HIV−TB) at a cut-off value of 5% and 13%, respectively. The corresponding ROC curves are shown in Figure 6C,D. On Gag-stimulated CD4+ T-cells of HIV+TB, the dominant activation profile was CD38+HLA-DR−Ki67−, similar to bulk CD4+ T-cells (results not shown) while CD8+ T-cells showed both CD38+HLA-DR+Ki67− and CD38+HLA-DR−Ki67− phenotypes (results not shown). When we evaluated the effect of TB treatment on the activation profile of ESAT-6/CFP-10-stimulated CD4+ T-cells, we observed significant declines in the proportion of CD38+HLA-DR+ CD4+ T-cells of HIV−TB ($p = 0.001$, Supplementary Figure S4A). In HIV+TB, only the CD38+HLA-DR+Ki67+ phenotype on ESAT-6/CFP-10 stimulated CD4+ T-cells declined at two months ($p = 0.04$, Supplementary Figure S4B). On Gag stimulated total cytokine+ CD8+ T-cells, we did not observe significant declines in the frequencies of cells expressing the CD38+HLA-DR+Ki67− ($p = 0.69$) or CD38+HLA-DR−Ki67− ($p = 0.78$) phenotypes (Supplementary Figure S4C).

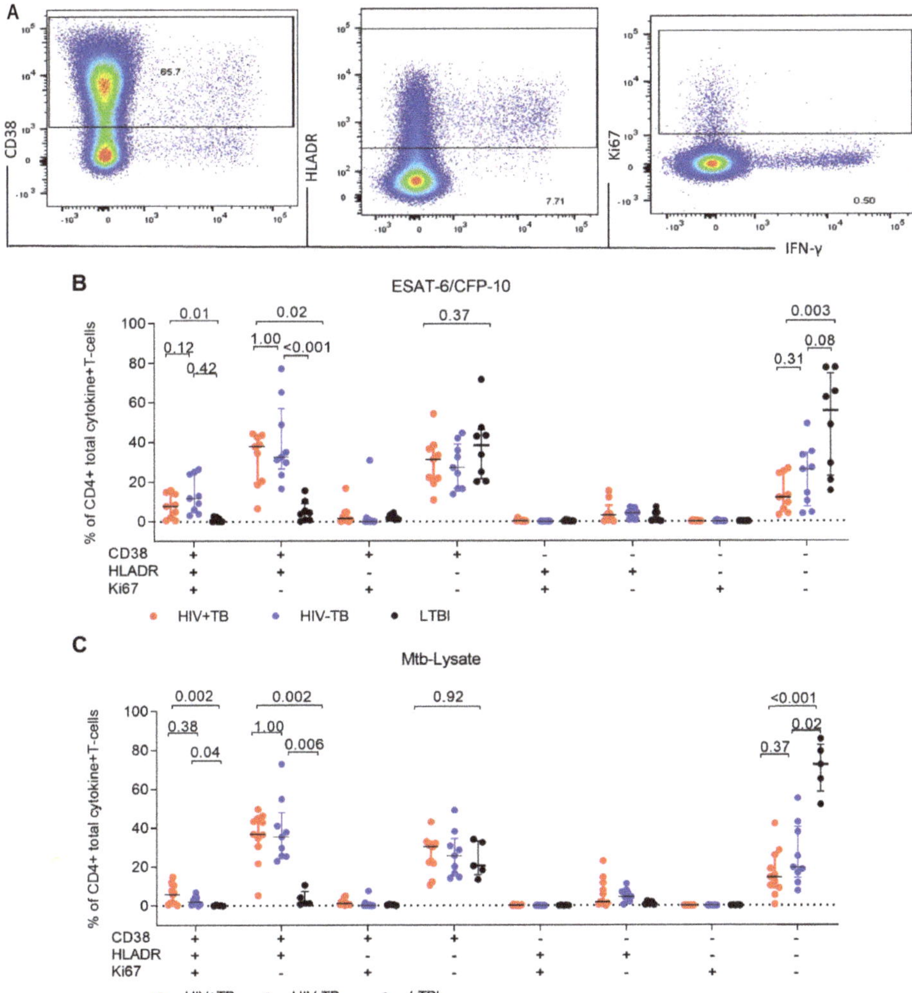

Figure 5. Comparison of the activation profiles of *Mtb*-specific CD4+ T-cells between HIV+TB, HIV−TB, and LTBI participants. (**A**) Representative dot plots of the activation profile on CD4+ T-cells of an HIV+TB patient in response to ESAT-6/CFP-10 stimulation. The proportion of each subset on ESAT-6/CFP-10 (**B**) and Mtb-Lysate (**C**) stimulated total cytokine+CD4+ T-cells. Statistical comparisons were made using the Kruskal–Wallis with Dunn's test for multiple comparisons. We used Boolean gates to derive all possible combinations of the co-expressed markers.

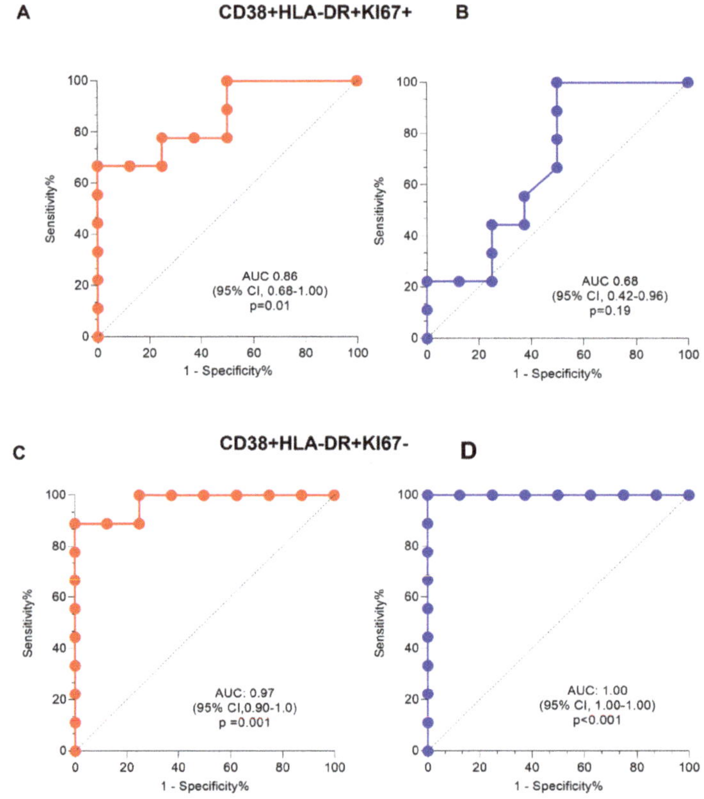

Figure 6. Receiver operating characteristics (ROC) curve analysis of the activation phenotype on total cytokine producing CD4+ T-cells for discriminating between TB and LTBI. Phenotypic analysis was only conducted in participants with a positive ESAT-6/CFP-10 response defined by a false discovery rate (FDR) < 0.05 using mixture models for single cell assays (MIMOSA) analysis: HIV+TB (n = 9); HIV−TB (n = 9); LTBI (n = 8). Comparison of the area under the curve for the phenotype CD38+HLA-DR+Ki67+ (top panel, **A**,**B**) and CD38+HLA-DR+Ki67- (bottom panel, **C**,**D**) in HIV+TB (red) and HIV−TB (blue). ROC curve analysis was done in Prism. The area under the curve (AUC) and p-values are shown. The dotted lines represent an AUC of 0.5.

2.4. A High CD27MFI Ratio Discriminates Active TB from Latent TB and Declines after Two Months of TB Treatment

To assess the effect of active disease on the memory phenotype of *Mtb*-specific CD4+ T-cells, we analyzed the Boolean combination of CD27, KLRG1, CD45RA, and CCR7 (Figure 7A shows representative dot plots). However, KLRG1 added no further information to the classification of memory subsets, therefore we used CD27, CD45RA, and CCR7 to define the memory phenotype. In agreement with previous reports, CD27 expression was downregulated on ESAT-6/CFP-10-stimulated CD4+ T-cells with an effector memory phenotype characterized by a high proportion of CD27−CD45RA−CCR7− within the total cytokine+ compartment ($p < 0.001$ for HIV−TB vs. LTBI and $p = 0.02$ for HIV+TB vs. LTBI, Figure 7B). In contrast, Gag-stimulated CD4+ T-cells were predominantly of a central memory phenotype (CD27+CD45RA−CCR7+; Supplementary Figure S5) with a dominant transitional memory phenotype (CD27+CD45RA+/−CCR7−; results not shown) observed on CD8+ T-cells. We did not observe statistically significant differences when we assessed the effect of TB treatment on the memory phenotype of ESAT-6/CFP-10-stimulated T-cells (Figure 7C,D).

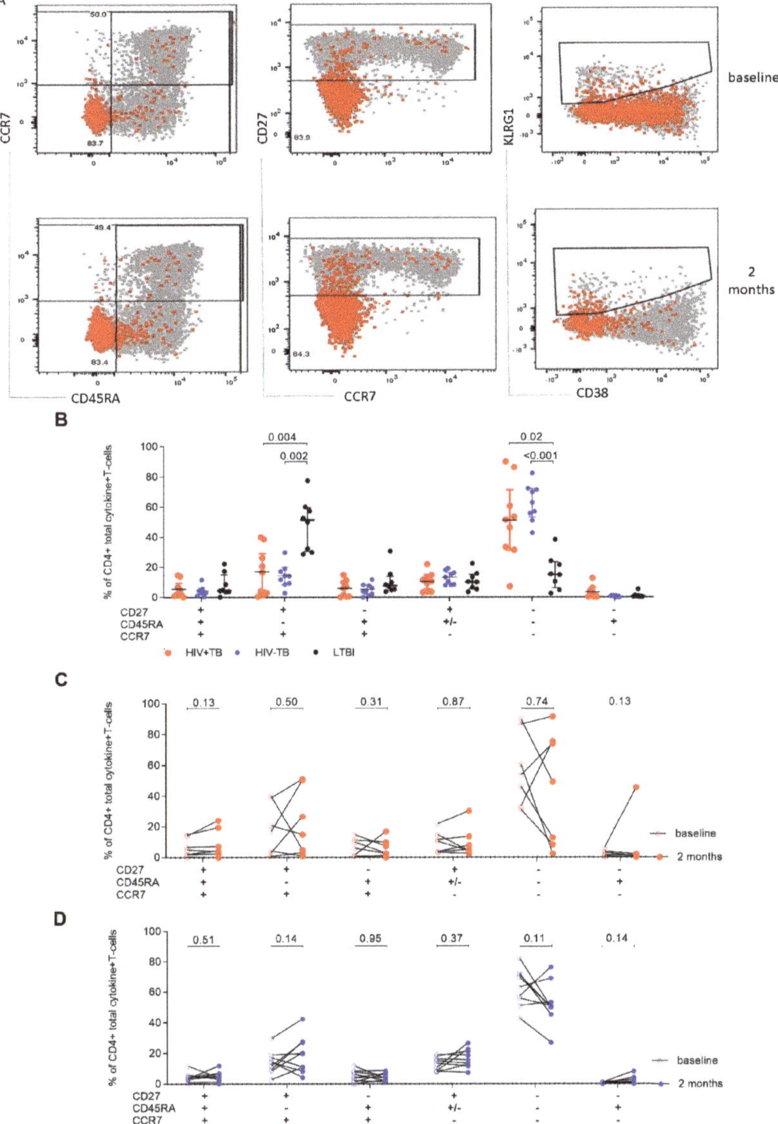

Figure 7. Comparison of the memory phenotype of ESAT-6/CFP-10-stimulated CD4+ T-cells between active TB and LTBI participants. (**A**) Representative plot of the total cytokine production in the memory compartment in response to ESAT-6/CFP-10 stimulation at baseline and two months (fu) timepoints. (**B**) Comparison of the memory phenotype of ESAT-6/CFP-10 stimulated total cytokine+CD4+cells between HIV+TB and HIV−TB. Changes in the memory phenotype between baseline and two months of TB treatment in HIV+TB (red, (**C**)) and HIV−TB (blue, (**D**)). Statistical comparisons were made using Kruskal–Wallis with Dunn's test for multiple comparisons. Differences between time points were assessed using the Wilcoxon signed rank test.

We next evaluated CD27 expression as a ratio of the MFI of CD27 on CD4+ T-cells to that of CD27 on total cytokine+ CD4+ T-cells, using a modification of the calculation as suggested by Portevin et al. [10]. We observed that a high ratio was associated with active TB on ESAT-6/CFP-10 ($p = 0.01$ for HIV−TB vs.

LTBI and $p = 0.05$ for HIV+TB vs. LTBI, Figure 8A), but not the Mtb-Lysate stimulated CD4+ T-cells ($p = 0.10$; results not shown). Significant declines in the ratio on ESAT-6/CFP-10-specific CD4+ T-cells were observed at two months of treatment in both HIV+TB ($p = 0.02$, Figure 8B) and HIV−TB ($p = 0.004$, Figure 8C). This decline was also observed on Gag stimulated CD8+T-cells of HIV+TB ($p = 0.03$, Figure 8D). We then assessed the diagnostic accuracy of using the CD27 MFI ratio on ESAT-6/CFP-10 stimulated CD4+ T-cells and found comparable sensitivity in HIV−TB (67%, 95% CI 35–88) and HIV+TB (63%, 95% CI 31–86) at a cut-off of 1.4. The corresponding ROC curves are shown in Figure 8E,F. Overall, our results suggest that an approach using the CD27MFI ratio on ESAT-6/CFP-10-specific total cytokine+ CD4+ T-cells could be used to discriminate between disease and infection in HIV-uninfected individuals and monitor treatment response during active TB, regardless of HIV status.

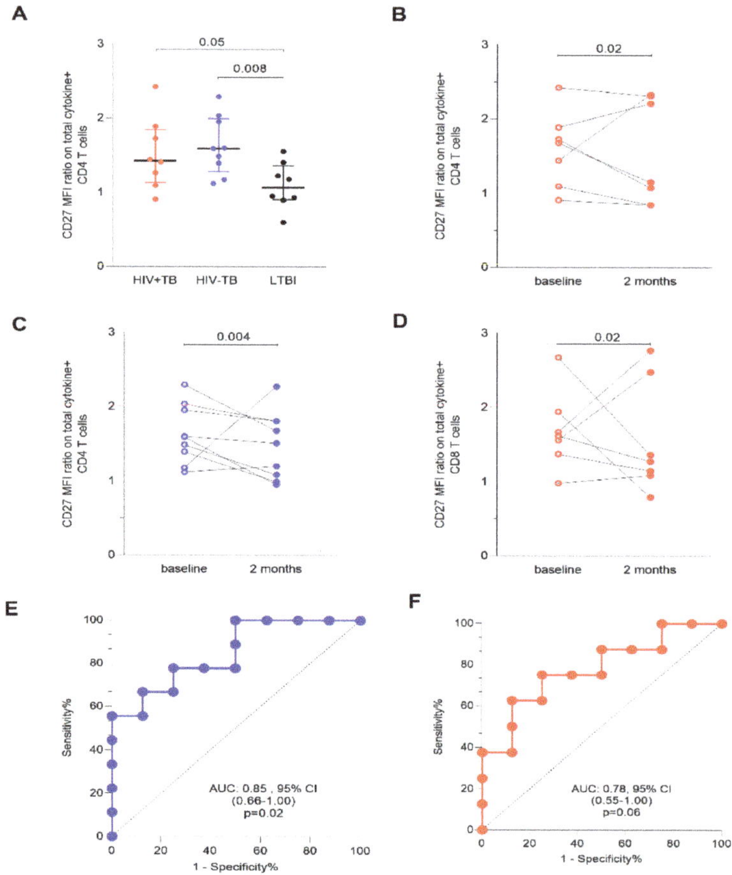

Figure 8. Sensitivity of the CD27MFI ratio on ESAT-6/CFP-10-specific CD4+ T-cells to detect active TB. (**A**) Comparison of the CD27MFI ratio between active TB and latent TB. CD27MFI ratio was calculated as the CD27MFI on bulk CD4 T-cells divided by CD27MFI on total cytokine+ CD4+ T-cells. Change in the CD27 MFI ratio of total cytokine+CD4+ T-cells between baseline and two months of TB treatment in (**B**) HIV+TB, (**C**) HIV−TB, and Gag-stimulated total cytokine+CD8+T-cells (**D**). Differences between time points were assessed using the Wilcoxon signed rank test. (**E**) and (**F**) show the receiver operating characteristics (ROC) curves of the CD27 MFI ratio in HIV−TB (blue) and HIV+TB (red), respectively, as assessed by the ROC curve analysis in Prism. The dotted lines represent an area under the curve (AUC) of 0.5.

3. Discussion

Our results show that the activation and memory phenotype of *Mtb*-specific CD4+ T-cells can be used to discriminate between active TB and LTBI as well as to monitor declines in mycobacterial burden after two months of ATT. These findings extend the results previously reported in ART naïve participants [12,13,20] to individuals on ART.

Conflicting reports on the assessment of the functional profile in HIV+TB have been published, reflecting the challenges of relying on cytokine profiles as biomarkers of TB. While some studies have shown that in active TB there is a depletion of single IFN-γ expression [21–23] accompanied by a dominant TNFα profile [24], others have reported that a polyfunctional profile is associated with HIV+TB cases [25]. We observed dominant single IFN-γ and dual functional IFN-γ+TNFα+ CD4+ T-cells in response to ESAT-6/CFP-10 and Mtb-Lysate stimulation consistent with reports from South African cohorts [12,13]. Whether these inconsistencies are due to differences in demographics of the study cohorts, study design, or flow cytometry technical differences is not very clear. More useful than the polyfunctional profile was the PFS that increased in response to ATT in HIV−TB, although there was no difference in scores between HIV+TB and HIV−TB. Therefore, using functional profiles as assessed by COMPASS may be a better measure of response to treatment. Stimulation with Mtb-Lysate also revealed a dominant single TNFα response that had not been evident on ESAT-6/CFP-10 CD4+ T-cells. In addition, all participants responded to Mtb-Lysate, which suggests that this antigen formulation may offer advantages for detecting responses in participants who are unresponsive to ESAT-6/CFP-10 [13]. This is because Mtb-Lysate detects mycobacterial responses induced by Mtb, BCG vaccination, and/or exposure to environmental mycobacteria, [13] which could be a drawback on the specificity of the responses obtained.

HIV is a major risk factor for TB, with deficiencies in *Mtb*-specific CD4+ T-cells possibly contributing to the risk of active TB prior to declines in peripheral blood CD4+ T-cells [26]. While ART-induced immune reconstitution is well described, the impact of ART on co-pathogen specific CD4+ T-cell responses is poorly understood [17,27]. For example, although ART induced declines in bulk CD4+ and CD8+ T-cells co-expressing CD38 and HLA-DR after a year of ART, the activation was still higher than on T-cells from HIV-uninfected individuals [18]. Since the activation level of T-cells is one of the factors influencing the extent of immune restoration [18], the use of blood-based markers may be an effective tool to assess the risk of active TB prior to ART initiation. In our cohort of participants who had been on ART for a median of one year, we found that the immune activation in HIV+TB, measured as a proportion of CD38+HLA-DR+totalcytokine+ ESAT-6/CFP-10-responding CD4+ T-cells, was comparable to HIV−TB but higher than LTBI at the time of TB diagnosis. In addition, significant declines in activation at two months post-ATT were observed only in HIV−TB. This could indicate that we needed a longer follow-up period and a larger sample size to observe a decline in response to TB treatment in the HIV+TB group.

Prior studies have shown that the activation profile of Mtb-specific CD4+ T-cells defined by CD38, HLA-DR, and Ki67 in combination [11–13] or alone [14] can distinguish latent from active TB with high sensitivity. Our results of a high sensitivity for the co-expression of CD38 and HLA-DR on *Mtb*-specific CD4+ T-cells as reflected by the ROC curve analysis agree with reports from Riou et al. [13] and Wilkinson et al. [12], in our analysis, the co-expression of CD38, HLA-DR, and Ki67 had poor sensitivity for TB diagnosis and the proportion of cells expressing this phenotype did not change after two months of TB treatment. The expression of Ki67 has also been shown to be a less robust marker for differentiating active disease from LTBI in a study of HIV infected participants who were not on ART [12].

The chronic antigen exposure that occurs in active TB disease drives the maturation of *Mtb*-specific CD4+ T-cells toward a late differentiated phenotype [28]. CD27 loss has been shown to differentiate active TB from LTBI [9,10,13,28] and associates with clinical disease severity and lung tissue damage in active TB [29]. Consistent with these reports, *Mtb*-specific CD4+ T-cells in response to ESAT-6/CFP-10 stimulation from active TB cases, regardless of HIV status, had a dominant effector memory phenotype compared to the central memory phenotype on LTBI. In contrast, Gag stimulated CD8+ T-cells had a dominant transitional memory phenotype that has been associated with HIV virologic control in participants not on ART and reduced immune activation [30].

Recently, a T-cell-activation marker-tuberculosis (TAM-TB) assay that measures the CD27 phenotype of IFN-γ producing cells was shown to have good performance characteristics in children [10]. We extended the analysis of the CD27 MFI ratio (measured as CD27 MFI on CD4+ T-cells divided by CD27 MFI on IFN-γ+CD4+ T-cells), suggested by Portevin et al. [10], to total cytokine+CD4+ T-cells, as recently reported [9]. We added new information regarding the CD27 MFI ratio in HIV+TB participants and showed that the ratio was highest on ESAT-6/CFP-10 responding total cytokine+ CD4+ T-cells in active TB participants regardless of HIV status. Previously, CD27 was analyzed as a % or MFI on IFN-γ+CD4+ T-cells of HIV+TB who were not on ART also showed good sensitivity in distinguishing active disease from infection [13]. In our analysis, TB treatment also caused declines in the CD27 MFI ratio in the HIV−TB participants. An increase in CD27 expression on *Mtb*-specific T-cells has also been shown to correlate with the conversion of sputum culture and lung repair measured as the reduction in lung tissue destruction on radiological examination at two months of ATT [29].

Our results should be interpreted in light of some limitations to the analyses. The sample size of our study groups was very small and was limited by studying participants with paired samples to detect early changes during ATT. This further reduced the sample sizes and may have possibly masked some differences and limited our ability to detect changes in certain phenotypes. Larger cohorts would be required to extend our findings. Whilst biomarkers for early treatment monitoring are important for identifying patients that are likely to respond well to shorter treatment regimens [31], the lack of an end of treatment timepoint and the fact that all patients had culture negative results at two months, limits our ability to comment on whether changes in the biomarkers are associated with favorable responses to ATT. Nonetheless, our results show that although TB treatment generally lasts six months, there are a number of patients that will culture convert by the end of the intensive phase of ATT. In addition, the inclusion of Gag stimulation allowed us to observe that these changes in T-cell activation appeared to be specific to the reduction of *Mtb* load during ATT because Gag-specific cells remained unaffected. However, the functional impact of the divergence in the memory phenotypes observed between Mtb and HIV-specific T-cells could not be determined due to the nature of the study design. Furthermore, we did not include HIV+ participants without TB as a control, which limits our ability to comment on the specificity of such responses.

Overall, our results show that CD38 and HLA-DR co-expression on *Mtb*-specific T-cells has good performance characteristics for active TB regardless of HIV status and has potential for early treatment monitoring in HIV+TB individuals on ART. The CD27 MFI ratio performed well on ESAT-6/CFP-10 responding total cytokine+ CD4+ T-cells in HIV−TB and needs further validation in large, well-powered cohorts to define its potential in monitoring early treatment responses in HIV+TB.

4. Materials and Methods

4.1. Study Participants

Participants with a diagnosis of pulmonary TB were recruited from 10 primary health clinics in Gaborone, Botswana during the period December 2017 to August 2018. Participants were recruited on the day of TB diagnosis or within five days thereof, prior to anti-TB treatment (ATT). All participants had a confirmed TB diagnosis by either smear microscopy or X-pert MTB/RIF (Cepheid, Inc., Sunnyvale, CA, USA). Sputum samples were collected at the time of diagnosis and cultured using Lowenstein–Jensen media, and phenotypic drug susceptibility testing was conducted. We collected samples from 10 healthy participants with no symptoms or previous history of TB, with a positive Quantiferon Gold Plus test (QIAGEN, Hilden, Germany) to serve as a comparison group. Blood samples were collected from all participants in sodium heparin tubes for peripheral blood mononuclear cell (PBMC) isolation. Participants with TB had study visits at 1- and 2-months post ATT, with sputum samples for culture and blood for PBMC processing collected at each visit. Samples for CD4 and viral load testing were collected at the baseline and two months and processed on the FACSCalibur (BD Biosciences, San Jose, CA, USA) and Abbott m2000sp/rt (Abbott, Chicago, IL, USA) instruments, respectively, at the ISO-accredited Botswana Harvard HIV Reference Laboratory (BHHRL).

4.2. Laboratory Methods

4.2.1. Cell Activation

PBMCs were isolated using the Ficoll–Hypaque density gradient method (GE Healthcare, Chicago, IL, USA) and stored in liquid nitrogen before shipping to the South African Tuberculosis Vaccine Initiative (SATVI) for processing. Cryopreserved cells were thawed and rested for two hours in Roswell Park Memorial Institute RPMI 1640 media containing 10% heat-inactivated fecal bovine serum (FBS) prior to antigen stimulation. We stimulated cells with a pool of early secretory antigen (ESAT-6) and culture filtrate protein (CFP-10, peptide pool referred to as ESCF) consisting of 17 and 16 peptides, respectively, overlapping by 10 amino acid sequences (1 μg/mL, GenScript, Piscataway, NJ, USA); *Mtb* cell lysate (MtbLy) ($H_{37}Rv$; 10 μg/mL, BEI Resources, Manassas, VA, USA), and HIV-1C Gag 15 mers overlapping by 10 amino acids (2 μg/mL, BEI Resources, Manassas, VA, USA). Phytohemagglutinin antigen (PHA; 2 μg/mL) was used as a positive control. A negative control without stimulation (NS) was also included. Stimulations were performed for two hours in a 37 °C CO_2 incubator, after which Brefeldin A (5 μg/mL, Sigma-Aldrich, St. Louis, MO, USA) was added and cells were further stimulated for four hours. All stimulations were conducted in the presence of the co-stimulatory antibodies, anti-CD28 and anti-CD49d (1 μg/mL, BD Biosciences, San Jose, CA, USA) for a total of six hours.

4.2.2. Cell Staining

After stimulations, cells were washed, stained with phycoerythrin (PE)-conjugated chemokine receptor antibody CCR7 (BD Biosciences, San Jose, CA, USA), and subsequently surface stained with the following antibodies: Brilliant violet (BV) 650-conjugated CD3 (BD Biosciences, San Jose, CA, USA); CD4-BV785 (BD); CD45RA-BV570 (BioLegend, San Diego, CA, USA); CD27–BV510 (BD Biosciences, San Jose, CA, USA); Fluorescein Isothiocynate (FITC)-conjugated HLA-DR (BioLegend, San Diego, CA, USA); Phycoerythrin (PECy5)-conjugated CD38 (BD Biosciences, San Jose, CA, USA); Peridinin-chlorophyll-protein complex (PerCP_eFluor710)-conjugated Killer Lectin-like receptor G (KLRG1, Invitrogen, Carlsbad, CA, USA); and LIVE-DEAD-Far Infrared (IR, Invitrogen, Shenzhen, China). The antibody panel including clone and catalog number in accordance with the MIFlowCyt guidelines [32] is shown in Supplementary Table S1. Cells were then fixed and permeabilized using Cytofix/Cytoperm buffer (BD Biosciences, San Jose, CA, USA) and stained with the antibodies Allophycocyanin (APC)-conjugated Interleukin (IL)-2 (BD Biosciences, San Jose, CA, USA); interferon gamma (IFN-γ)-AlexaFluor700 (BD Biosciences, San Jose, CA, USA); tumor necrosis factor (TNF)-α -PeCy7 (BioLegend, San Diego, CA, USA); and Ki67-BV421 (BD Biosciences, San Jose, CA, USA). Cells were fixed in 1% paraformaldehyde and acquired on a BD LSRII flow cytometer within one hour of fixation at the South African Tuberculosis Vaccine Initiative (SATVI). FACS data were analyzed in FlowJo (BD Biosciences, San Jose, CA, USA) and Boolean combination gating was used to calculate frequencies corresponding to eight and sixteen different combinations of cytokines and memory markers, respectively. The gating strategy is shown in Supplementary Figure S1.

4.3. Statistical Analysis

In order to compare the functional, activation, and memory profile of *Mtb*-specific CD4+ T-cells between antigen conditions or patient groups, we used the Wilcoxon rank sum, and Kruskal–Wallis with Dunn's tests for multiple comparisons as appropriate. Cytokine combinations were assessed using SPICE (simplified presentation of incredibly complex evaluations) software and data are reported after background subtraction [33]. Differences in the functional, activation, and memory profile of

Mtb-specific CD4+ T-cells between baseline and two months were analyzed using the Wilcoxon signed rank test. Receiver operating characteristics (ROC) curve analysis was used for single marker analysis. COMPASS (combinatorial polyfunctionality analysis of antigen-specific T-cells) and MIMOSA (mixture models for single cell assays) analysis was conducted in R (v1.1.463, The R Foundation for Statistical Computing, Vienna, Austria) using the packages as described previously [19]. To reliably measure the expression of phenotypic markers on antigen specific cells, we first determined whether the immune response to antigen stimulation was significantly higher when compared to non-specific cytokine production detectable in unstimulated samples. To do this, we applied MIMOSA analysis and defined antigen responders as participants with a false discovery rate (FDR) < 0.05 from this analysis. COMPASS utilizes a Bayesian hierarchical mixture model to identify antigen-specific changes simultaneously across all possible T-cell subsets [19]. Such an approach has a higher sensitivity and specificity for detecting true responses over alternative approaches such as basic log fold change [19]. A Markov Chain Monte Carlo algorithm computes posterior probabilities resulting in two scores; functionality and polyfunctionality that can be correlated with any outcome of interest. We analyzed data using STATA (v14.2; StataCorp, College Station, TX, USA) and GraphPad Prism (v8.1; GraphPad Software, La Jolla, CA, USA). A p-value < 0.05 was considered statistically significant and we adjusted for multiple comparisons where needed.

4.4. Ethical Considerations

This work received ethical approval from The University of Botswana IRB and the Human Research Development Committee (IRB of the Botswana Ministry of Health and Wellness, No: HPDME 13/18/1). All participants provided written informed consent.

Supplementary Materials: The following are available online at http://www.mdpi.com/2076-0817/9/3/180/s1, Figure S1: Gating Strategy, Figure S2: COMPASS analysis of Gag and Mtb-Lysate-stimulated CD8 and CD4 T-cells, Figure S3: Comparison of TNFα expression within total cytokine+CD4 T-cells, Figure S4: Comparison of activation profile of Gag-stimulated total cytokine+CD8+T-cells, Figure S5: Comparison of memory phenotypes between Mtb and HIV specific total cytokine+CD4+T-cells, Table S1: Antibody Panel.

Author Contributions: Conceptualization, S.G. and N.Z.; Methodology, T.R., C.A.M.M., E.N., T.J.S., and L.M.; Validation, C.A.M.M., L.M., E.N., and T.J.S.; Formal analysis, L.M.; Investigation, L.M.; Resources, S.G., S.S.S., T.J.S., and N.Z.; Data curation, L.M.; Writing—original draft preparation, L.M.; Writing—review and editing, T.R., S.M., I.K., T.M., S.G., S.S.S., E.N., and T.J.S.; Supervision, S.G., S.M., I.K., T.J.S., and E.N.; Project administration, S.G. and R.M.M.; Funding acquisition, S.G. All authors have read and agreed to the published version of the manuscript.

Funding: This work was supported by the sub-Saharan African Network for TB/HIV Research Excellence (SANTHE), a DELTAS Africa Initiative (grant # DEL-15-006 to L.M., S.M., T.M., and S.G.), the Fogarty International Center, National Institute of Mental Health and National Institute of Allergy and Infectious Diseases of the National Institutes of Health (grant # D43 TW010543 to S.M. and grant # K01AI118559 to SSS). The DELTAS Africa Initiative is an independent funding scheme of the African Academy of Sciences (AAS)'s Alliance for Accelerating Excellence in Science in Africa (AESA) and supported by the New Partnership for Africa's Development Planning and Coordinating Agency (NEPAD Agency) with funding from the Wellcome Trust (grant #107752/Z/15/Z) and the government of the United Kingdom. R.M. and T.M. were partially supported by the EDCTP2 program Trials of Excellence in Southern Africa (TESA II), supported under Horizon 2020, the European Union. The views expressed in this publication are those of the authors and not necessarily those of the AAS, AESA, NEPAD Agency, Wellcome Trust, National Institutes of Health, or the UK government. The funders had no role in the study design, data collection, and decision to publish or in the preparation of the manuscript.

Acknowledgments: We wish to thank One Dintwe from the Cape Town HVTN Immunology Laboratory (CHIL) for the kind donation of the HIV Gag peptides used for the intracellular cytokine-staining assay.

Conflicts of Interest: The authors declare no conflicts of interest.

References

1. WHO. Global TB Report. 2019. Available online: https://www.who.int/tb/publications/global_report/en/ (accessed on 17 October 2019).
2. Getahun, H.; Gunneberg, C.; Granich, R.; Nunn, P. HIV infection-associated tuberculosis: The epidemiology and the response. *Clin. Infect. Dis.* **2010**, *3*, S201–S207. [CrossRef] [PubMed]

3. Dye, C.; Williams, B.G. Tuberculosis decline in populations affected by HIV: A retrospective study of 12 countries in the WHO African Region. *Bull. World Health Organ.* **2019**, *97*, 405–414. [CrossRef] [PubMed]
4. Gupta, R.K.; Rice, B.; Brown, A.E.; Thomas, H.L.; Zenner, D.; Anderson, L.; Pedrazzoli, D.; Pozniak, A.; Abubakar, I.; Delpech, V.; et al. Does antiretroviral therapy reduce HIV-associated tuberculosis incidence to background rates? A national observational cohort study from England, Wales, and Northern Ireland. *Lancet HIV* **2015**, *2*, e243–e251. [CrossRef]
5. Gupta, A.; Wood, R.; Kaplan, R.; Bekker, L.G.; Lawn, S.D. Tuberculosis incidence rates during 8 years of follow-up of an antiretroviral treatment cohort in South Africa: Comparison with rates in the community. *PLoS ONE* **2012**, *7*, e34156. [CrossRef] [PubMed]
6. Geldmacher, C.; Zumla, A.; Hoelscher, M. Interaction between HIV and *Mycobacterium tuberculosis*: HIV-1-induced CD4 T-cell depletion and the development of active tuberculosis. *Cur. Opin. HIV AIDS* **2012**, *7*, 268–275. [CrossRef]
7. Geldmacher, C.; Schuetz, A.; Ngwenyama, N.; Casazza, J.P.; Sanga, E.; Saathoff, E.; Boehme, C.; Geis, S.; Maboko, L.; Singh, M.; et al. Early depletion of *Mycobacterium tuberculosis*-specific T helper 1 cell responses after HIV-1 infection. *J. Infect. Dis.* **2008**, *198*, 1590–1598. [CrossRef]
8. Tonaco, M.M.; Moreira, J.D.; Nunes, F.F.C.; Loures, C.M.G.; Souza, L.R.; Martins, J.M.; Silva, H.R.; Porto, A.H.R.; Toledo, V.; Miranda, S.S.; et al. Evaluation of profile and functionality of memory T cells in pulmonary tuberculosis. *Immunol. Lett.* **2017**, *192*, 52–60. [CrossRef]
9. Riou, C.; Tanko, R.F.; Soares, A.P.; Masson, L.; Werner, L.; Garrett, N.J.; Samsunder, N.; Karim, Q.A.; Karim, S.S.A.; Burgers, W.A. Restoration of CD4+ responses to copathogens in HIV-infected individuals on antiretroviral therapy is dependent on t cell memory phenotype. *J. Immunol.* **2015**, *195*, 2273–2281. [CrossRef]
10. Latorre, I.; Fernandez-Sanmartin, M.A.; Muriel-Moreno, B.; Villar-Hernandez, R.; Vila, S.; Souza-Galvao, M.L.; Stojanovic, Z.; Jimenez-Fuentes, M.A.; Centeno, C.; Ruiz-Manzano, J.; et al. Study of CD27 and CCR4 markers on specific CD4(+) T-Cells as immune tools for active and latent tuberculosis management. *Front. Immunol.* **2018**, *9*, 3094. [CrossRef]
11. Portevin, D.; Moukambi, F.; Clowes, P.; Bauer, A.; Chachage, M.; Ntinginya, N.E.; Mfinanga, E.; Said, K.; Haraka, F.; Rachow, A.; et al. Assessment of the novel *t*-cell activation marker-tuberculosis assay for diagnosis of active tuberculosis in children: A prospective proof-of-concept study. *Lancet Infect. Dis.* **2014**, *14*, 931–938. [CrossRef]
12. Adekambi, T.; Ibegbu, C.C.; Cagle, S.; Kalokhe, A.S.; Wang, Y.F.; Hu, Y.; Day, C.L.; Ray, S.M.; Rengarajan, J. Biomarkers on patient T cells diagnose active tuberculosis and monitor treatment response. *J. Clin. Investig.* **2015**, *125*, 1827–1838. [CrossRef]
13. Wilkinson, K.A.; Oni, T.; Gideon, H.P.; Goliath, R.; Wilkinson, R.J.; Riou, C. Activation Profile of *Mycobacterium tuberculosis*-Specific CD4(+) T Cells Reflects Disease Activity Irrespective of HIV Status. *Am. J. Respir. Crit. Care Med.* **2016**, *193*, 1307–1310. [CrossRef] [PubMed]
14. Riou, C.; Berkowitz, N.; Goliath, R.; Burgers, W.A.; Wilkinson, R.J. Analysis of the phenotype of *Mycobacterium tuberculosis*-specific CD4+ T cells to discriminate latent from active tuberculosis in HIV-uninfected and HIV-infected individuals. *Front. Immunol.* **2017**, *8*, 968. [CrossRef] [PubMed]
15. Musvosvi, M.; Duffy, D.; Filander, E.; Africa, H.; Mabwe, S.; Jaxa, L.; Bilek, N.; Llibre, A.; Rouilly, V.; Hatherill, M.; et al. T-cell biomarkers for diagnosis of tuberculosis: Candidate evaluation by a simple whole blood assay for clinical translation. *Eur. Respir. J. Off. J. Eur. Soc. Clin. Respir. Physiol.* **2018**, *51*, 1800153. [CrossRef] [PubMed]
16. Chun, T.-W.; Justement, J.S.; Sanford, C.; Hallahan, C.W.; Planta, M.A.; Loutfy, M.; Kottilil, S.; Moir, S.; Kovacs, C.; Fauci, A.S. Relationship between the frequency of HIV-specific CD8+T cells and the level of CD38+CD8+T cells in untreated HIV-infected individuals. *Proc. Nat. Acad. Sci. USA* **2004**, *101*, 2464–2469. [CrossRef]
17. Fletcher, H.A.; Snowden, M.A.; Landry, B.; Rida, W.; Satti, I.; Harris, S.A.; Matsumiya, M.; Tanner, R.; O'Shea, M.K.; Dheenadhayalan, V.; et al. T-cell activation is an immune correlate of risk in BCG vaccinated infants. *Nat. Commun.* **2016**, *7*, 11290. [CrossRef]
18. Tanko, R.F.; Soares, A.P.; Masson, L.; Garrett, N.J.; Samsunder, N.; Karim, Q.A.; Karim, S.S.A.; Riou, C.; Burgers, W.A. Residual T cell activation and skewed CD8+ T cell memory differentiation despite antiretroviral therapy-induced HIV suppression. *Clin. Immunol.* **2018**, *195*, 127–138. [CrossRef]

19. Lin, L.; Finak, G.; Ushey, K.; Seshadri, C.; Hawn, T.R.; Frahm, N.; Scriba, T.J.; Mahomed, H.; Hanekom, W.; Bart, P.-A.; et al. COMPASS identifies T-cell subsets correlated with clinical outcomes. *Nat. Biotechnol.* **2015**, *33*, 610–616. [CrossRef]
20. Schuetz, A.; Haule, A.; Reither, K.; Ngwenyama, N.; Rachow, A.; Meyerhans, A.; Maboko, L.; Koup, R.A.; Hoelscher, M.; Geldmacher, C. Monitoring CD27 expression to evaluate Mycobacterium tuberculosis activity in HIV-1 infected individuals in vivo. *PLoS ONE* **2011**, *6*, e27284. [CrossRef]
21. Riou, C.; Bunjun, R.; Muller, T.L.; Kiravu, A.; Ginbot, Z.; Oni, T.; Goliath, R.; Wilkinson, R.J.; Burgers, W.A. Selective reduction of IFN-gamma single positive mycobacteria-specific CD4+ T cells in HIV-1 infected individuals with latent tuberculosis infection. *Tuberculosis* **2016**, *101*, 25–30. [CrossRef]
22. Sutherland, R.; Yang, H.; Scriba, T.J.; Ondondo, B.; Robinson, N.; Conlon, C.; Suttill, A.; McShane, H.; Fidler, S.; McMichael, A.; et al. Impaired IFN-gamma-secreting capacity in mycobacterial antigen-specific CD4 T cells during chronic HIV-1 infection despite long-term HAART. *AIDS* **2006**, *20*, 821–829. [CrossRef] [PubMed]
23. Liang, L.; Shi, R.; Liu, X.; Yuan, X.; Zheng, S.; Zhang, G.; Wang, W.; Wang, J.; England, K.; Via, L.E.; et al. Interferon-gamma response to the treatment of active pulmonary and extra-pulmonary tuberculosis. *Int. J. Tuberc. Lung Dis* **2017**, *21*, 1145–1149. [CrossRef] [PubMed]
24. Harari, A.; Rozot, V.; Enders, F.B.; Perreau, M.; Stalder, J.M.; Nicod, L.P.; Cavassini, M.; Calandra, T.; Blanchet, C.L.; Jaton, K.; et al. Dominant TNF-alpha+ *Mycobacterium tuberculosis*-specific CD4+ T cell responses discriminate between latent infection and active disease. *Nat. Med.* **2011**, *17*, 372–376. [CrossRef] [PubMed]
25. Chiacchio, T.; Petruccioli, E.; Vanini, V.; Cuzzi, G.; Pinnetti, C.; Sampaolesi, A.; Antinori, A.; Girardi, E.; Goletti, D. Polyfunctional T-cells and effector memory phenotype are associated with active TB in HIV-infected patients. *J. Infect.* **2014**, *69*, 533–545. [CrossRef] [PubMed]
26. Geldmacher, C.; Koup, R.A. Pathogen-specific T cell depletion and reactivation of opportunistic pathogens in HIV infection. *Trends Immunol.* **2012**, *33*, 207–214. [CrossRef] [PubMed]
27. Chiacchio, T.; Petruccioli, E.; Vanini, V.; Cuzzi, G.; La Manna, M.P.; Orlando, V.; Pinnetti, C.; Sampaolesi, A.; Antinori, A.; Caccamo, N.; et al. Impact of antiretroviral and tuberculosis therapies on CD4+ and CD8+ HIV/M. tuberculosis-specific T-cell in co-infected subjects. *Immunol. Lett.* **2018**, *198*, 33–43. [CrossRef]
28. Ahmed, M.I.M.; Ntinginya, N.E.; Kibiki, G.; Mtafya, B.A.; Semvua, H.; Mpagama, S.; Mtabho, C.; Saathoff, E.; Held, K.; Loose, R.; et al. Phenotypic Changes on Mycobacterium Tuberculosis-Specific CD4 T Cells as Surrogate Markers for Tuberculosis Treatment Efficacy. *Front. Immunol.* **2018**, *9*, 2247. [CrossRef]
29. Saharia, K.K.; Petrovas, C.; Ferrando-Martinez, S.; Leal, M.; Luque, R.; Ive, P.; Luetkemeyer, A.; Havlir, D.; Koup, R.A. Tuberculosis therapy modifies the cytokine profile, maturation state, and expression of inhibitory molecules on Mycobacterium tuberculosis-Specific CD4+ T-Cells. *PLoS ONE* **2016**, *11*, e0158262. [CrossRef]
30. Nikitina, I.Y.; Kondratuk, N.A.; Kosmiadi, G.A.; Amansahedov, R.B.; Vasilyeva, I.A.; Ganusov, V.V.; Lyadova, I.V. Mtb-Specific CD27low CD4 T Cells as markers of lung tissue destruction during pulmonary tuberculosis in humans. *PLoS ONE* **2012**, *7*, e43733. [CrossRef]
31. Burgers, W.A.; Riou, C.; Mlotshwa, M.; Maenetje, P.; De Assis Rosa, D.; Brenchley, J.; Mlisana, K.; Douek, D.C.; Koup, R.; Roederer, M.; et al. Association of HIV-specific and total CD8+ T memory phenotypes in subtype C HIV-1 infection with viral set point. *J. Immunol* **2009**, *182*, 4751–4761. [CrossRef]
32. Lee, J.A.; Spidlen, J.; Boyce, K.; Cai, J.; Crosbie, N.; Dalphin, M.; Furlong, J.; Gasparetto, M.; Goldberg, M.; Goralczyk, E.M.; et al. MIFlowCyt: The minimum information about a flow cytometry experiment. *Cytom. Part A J. Int. Soc. Anal. Cytol.* **2008**, *73*, 926–930. [CrossRef] [PubMed]
33. Roederer, M.; Nozzi, J.L.; Nason, M.C. SPICE: Exploration and analysis of post-cytometric complex multivariate datasets. *Cytom. Part A J. Int. Soc. Anal. Cytol.* **2011**, *79*, 167–174. [CrossRef] [PubMed]

Article

MmpS5-MmpL5 Transporters Provide *Mycobacterium smegmatis* Resistance to imidazo[1,2-*b*][1,2,4,5]tetrazines

Dmitry A. Maslov *, Kirill V. Shur, Aleksey A. Vatlin and Valery N. Danilenko

Laboratory of Bacterial Genetics, Vavilov Institute of General Genetics, Russian Academy of Sciences, Moscow 119333, Russia; shurkirill@gmail.com (K.V.S.); vatlin_alexey123@mail.ru (A.A.V.); valerid@vigg.ru (V.N.D.)

* Correspondence: Maslov_da@vigg.ru

Received: 4 February 2020; Accepted: 25 February 2020; Published: 28 February 2020

Abstract: The emergence and spread of drug-resistant *Mycobacterium tuberculosis* strains (including MDR, XDR, and TDR) force scientists worldwide to search for new anti-tuberculosis drugs. We have previously reported a number of imidazo[1,2-*b*][1,2,4,5]tetrazines–putative inhibitors of mycobacterial eukaryotic-type serine-threonine protein-kinases, active against *M. tuberculosis*. Whole genomic sequences of spontaneous drug-resistant *M. smegmatis* mutants revealed four genes possibly involved in imidazo[1,2-*b*][1,2,4,5]tetrazines resistance; however, the exact mechanism of resistance remain unknown. We used different approaches (construction of targeted mutants, overexpression of the wild-type (*w.t.*) and mutant genes, and gene-expression studies) to assess the role of the previously identified mutations. We show that mutations in *MSMEG_1380* gene lead to overexpression of the *mmpS5-mmpL5* operon in *M. smegmatis*, thus providing resistance to imidazo[1,2-*b*][1,2,4,5]tetrazines by increased efflux through the MmpS5-MmpL5 system, similarly to the mechanisms of resistance described for *M. tuberculosis* and *M. abscessus*. Mycobacterial MmpS5-MmpL5 transporters should be considered as an MDR-efflux system and they should be taken into account at early stages of anti-tuberculosis drug development.

Keywords: *Mycobacterium smegmatis*; imidazo[1,2-*b*][1,2,4,5]tetrazine; MmpS5-MmpL5; efflux; drug discovery; drug resistance; tuberculosis

---

1. Introduction

Tuberculosis (TB), caused by *Mycobacterium tuberculosis*, is currently the leading killer among the infectious diseases caused by one infectious agent, responsible for an estimate of 1.2 million deaths in 2018 [1]. According to WHO, 1.7 billion people globally are infected with *M. tuberculosis*, and thus are at risk of developing the disease [1]. The emergence and spread of multidrug resistant TB (MDR-TB, defined as TB resistant to rifampicin and isoniazid), extensively drug-resistant TB (defined as MDR-TB with resistance to the fluoroquinolones and second-line injectables) and totally drug-resistant TB (TDR-TB) is a global threat to world-wide TB control [2–5]. Long treatment times (six months for drug-susceptible TB and up to two years for DR-TB) often lead to bad patient compliance, which is one of the causes of drug resistance development and results in worse treatment outcomes. Thus, researchers are forced to search for novel anti-TB drugs and shorter regimens [6].

Eukaryotic-type serine-threonine protein-kinases (ESTPKs) play a key role in *M. tuberculosis* life cycle regulation, controlling some of its vital aspects such as cell division and survival within host macrophages, and, therefore, they represent attractive targets for drug development [7,8]. We have previously described a number of imidazo[1,2-*b*][1,2,4,5]tetrazines with a promising antibacterial activity on *M. tuberculosis* and *M. smegmatis* [9]. Most of these compounds showed activity as potential

ESTPK inhibitors in the original *M. smegmatis aphVIII+* test-system [10,11], and two of them were able to bind to the *M. tuberculosis* PknB adenine-binding pocket according to docking studies [10]. Despite the predicted activity as ESTPK-inhibitors, both the exact mechanism of action and the mechanism of resistance to these compounds are still unknown.

We were able to obtain spontaneous *M. smegmatis* mutants resistant to four imidazo[1,2-*b*][1,2,4,5] tetrazines (**3a**, **3c**, **3h** and **3n**, Figure 1), which had cross-resistance among them, suggesting a common mechanism of drug-resistance [9]. Whole-genomic sequencing and comparative genomic analysis revealed mutations in *MSMEG_0641* (binding-protein-dependent transporters inner membrane component) in 1 mutant, in *MSMEG_1601* (hypothetical protein) in seven mutants, in *MSMEG_2087* (transporter small conductance mechanosensitive ion channel (MscS) family protein) in one mutant [12], while all the mutants carried different mutations in *MSMEG_1380* (AcrR/TetR_N transcriptional regulator) – 1 nonsynonymous SNP, 2 insertions leading to a frameshift, 2 duplications (6 and 501 base pairs-long) and one deletion [13].

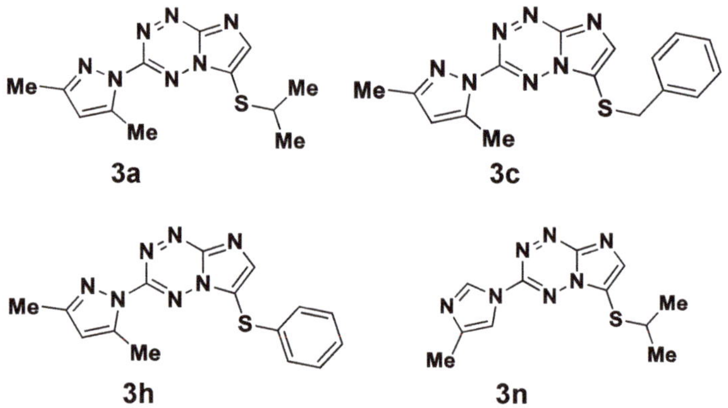

Figure 1. Chemical structures of imidazo[1,2-*b*][1,2,4,5]tetrazines [9].

In this article we describe the investigation of these mutations' role in mycobacterial drug resistance to imidazo[1,2-*b*][1,2,4,5]tetrazines by different approaches: construction of targeted mutants, overexpression of the wild-type (*w.t.*) and mutant genes, and gene-expression studies.

2. Results

2.1. Mutations in MSMEG_1380 Gene Lead to Imidazo[1,2-b][1,2,4,5]tetrazines Resistance in M. smegmatis

The list of nonsynonymous mutations found in spontaneous drug-resistant *M. smegmatis* mutants used in this study is presented in Table 1. We were able to construct targeted *M. smegmatis* mutants harboring each mutation in genes *MSMEG_0641*, *MSMEG_1601*, and *MSMEG_2087*, as well as five mutations found in *MSMEG_1380* gene using the p2NIL/pGOAL19 suicide system [14] for homologous recombination (Table 1).

We examined the minimal inhibitory concentrations (MICs) of the four compounds on the recombinant *M. smegmatis* strains and found that mutations in *MSMEG_1380* gene lead to elevated MICs as compared to the *w.t.* strain (4×MIC for the compound **3a**, at least 2×MIC for the compound **3c**, at least 2×MIC for the compound **3h**, and 4×MIC for the compound **3n**), while recombinants harboring mutations in genes *MSMEG_0641*, *MSMEG_1601*, and *MSMEG_2087* had the same MICs as the *w.t.* strain (Table 2).

Table 1. Bacterial strains used in the study.

Bacterial Strains		
Name	**Comment**	**Origin**
M. smegmatis mc2 155	Wild-type (*w.t.*) strain	
M. smegmatis atR1	Spontaneous mutant of *mc2 155*. Mutations: $Y_{52}H$ (TAC>CAC) in *MSMEG_1601*; del LLA_{41-43} (del $GCTGCTCGC_{480-488}$) in *MSMEG_1380*.	[9]
M. smegmatis atR2	Spontaneous mutant of *mc2 155*. Mutations: $Y_{52}H$ (TAC>CAC) in *MSMEG_1601*; ins $GC_{425-426}$ (frameshift) in *MSMEG_1380*.	[9]
M. smegmatis atR9	Spontaneous mutant of *mc2 155*. Mutations: $Y_{188}C$ (TAC>TGC) in *MSMEG_2087*; ins C_8 (frameshift) in *MSMEG_1380*.	[9]
M. smegmatis atR10	Spontaneous mutant of *mc2 155*. Mutations: $R_{233}S$ (CGT>AGT) in *MSMEG_0641*; ins C_8 (frameshift) in *MSMEG_1380*.	[9]
M. smegmatis atR14	Spontaneous mutant of *mc2 155*. Mutations: $Y_{52}H$ (TAC>CAC) in *MSMEG_1601*; ins G_{448} (frameshift) in *MSMEG_1380*.	[9]
M. smegmatis atR19	Spontaneous mutant of *mc2 155*. Mutations: $Y_{52}H$ (TAC>CAC) in *MSMEG_1601*; $T_{52}V$ (ACG>GTG) in *MSMEG_1380*.	[9]
M. smegmatis atR33	Spontaneous mutant of *mc2 155*. Mutations: ins VG_{52-53} (ins $GTGGGC_{154-159}$) in *MSMEG_1380*.	[9]
M. smegmatis atR37	Spontaneous mutant of *mc2 155*. Mutation: del C 662 (frameshift) in *MSMEG_1380*.	[9]
M. smegmatis atR1c	Recombinant strain, mutation: del LLA_{41-43} (del $GCTGCTCGC_{480-488}$) in *MSMEG_1380*.	This study
M. smegmatis atR2c	Recombinant strain, mutation: ins $GC_{425-426}$ (frameshift) in *MSMEG_1380*.	This study
M. smegmatis atR9c	Recombinant strain, mutation: ins C_8 (frameshift) in *MSMEG_1380*.	This study
M. smegmatis atR14c	Recombinant strain, mutation: ins G_{448} (frameshift) in *MSMEG_1380*.	This study
M. smegmatis atR33c	Recombinant strain, mutation: ins VG_{52-53} (ins $GTGGGC_{154-159}$) in *MSMEG_1380*	This study
M. smegmatis 0641c	Recombinant strain, mutation: $R_{233}S$ (CGT>AGT) in *MSMEG_0641*.	This study
M. smegmatis 1601c	Recombinant strain, mutation: $Y_{52}H$ (TAC>CAC) in *MSMEG_1601*	This study
M. smegmatis 2087c	Recombinant strain, mutation: $Y_{188}C$ (TAC>TGC) in *MSMEG_2087*	This study

Table 2. Imidazo[1,2-*b*][1,2,4,5]tetrazines MICs on *M. smegmatis* strains in liquid medium.

Compound	***M. smegmatis* Strains MICs, μg/mL**								
	mc2 155	*atR1c*	*atR2c*	*atR9c*	*atR14c*	*atR33c*	*0641c*	*1601c*	*2087c*
3a	128	512	512	512	512	512	128	128	128
3c	64	>128 *	>128 *	>128 *	>128 *	>128 *	64	64	64
3h	128	>256 *	>256 *	>256 *	>256 *	>256 *	128	128	128
3n	64	256	256	256	256	256	64	64	64

* The compounds were not soluble at higher concentrations; bacterial growth was observed at the stated concentrations.

Thus we have shown that only the mutations in *MSMEG_1380* are responsible for imidazo[1,2-*b*][1,2,4,5]tetrazines resistance in *M. smegmatis*.

2.2. W.t. MSMEG_1380 Overexpression Increases M. smegmatis Susceptibility to imidazo[1,2-b][1,2,4,5]tetrazines

In order to investigate further the role of *MSMEG_1380* in *M. smegmatis* resistance to imidazo[1,2-*b*][1,2,4,5]tetrazines, we cloned the *w.t. MSMEG_1380* gene and two of its mutant variants in the tetracycline inducible plasmid pMINDKm⁻ [15].

We used the paper-disc assay to assess the drug susceptibility of *M. smegmatis* strains to imidazo[1,2-*b*][1,2,4,5]tetrazines and found that the overexpression of the *w.t. MSMEG_1380* gene increases *M. smegmatis* susceptibility to the tested compounds, while the overexpression of its mutant variants had no effect on the phenotype (Table 3), thus suggesting that the disruption of MSMEG_1380 protein's function leads to the drug-resistant phenotype.

Table 3. Growth inhibition halos, produced by imidazo[1,2-*b*][1,2,4,5]tetrazines on *M. smegmatis* strains.

Compound	Concentration, nmole/disc	Growth Inhibition Halo, mm			
		M. smegmatis Transformants			
		pMINDKm⁻	pMINDKm⁻:*msmeg_1380*	pMINDKm⁻:*msmeg_1380-19*	pMINDKm⁻:*msmeg_1380-33*
3a	300	9.8 ± 1.5	17.0 ± 3.6	8.8 ± 0.8	8.2 ± 0.6
3c	300	7.0 ± 0.8	15.0 ± 2.9	6.3 ± 0.5	6.3 ± 0.5
3h	40	6.7 ± 0.5	11.5 ± 0.4	6.3 ± 0.5	6.5 ± 0.4
3n	100	9.7 ± 2.4	16.0 ± 0.8	9.2 ± 1.3	9.3 ± 1.2

2.3. *MSMEG_1380 Represses the Expression of the mmpS5-mmpL5 Operon in M. smegmatis*

MSMEG_1380 gene lies 179 b.p. upstream the *mmpS5-mmpL5* operon (genes *MSMEG_1381-MSMEG_1382*) in the *M. smegmatis* genome and is transcribed in the opposite direction (Figure 2). The structure of this operon is conserved in different mycobacterial species: the *mmpS5-mmpL5* genes are controlled by a TetR-family transcriptional repressor, encoded by a gene located upstream the operon. Mutations in genes encoding the TetR-repressor, which lead to the upregulation of the *mmpS5-mmpL5* genes, are involved in *M. abscessus* resistance to the derivatives of thiacetazone [16], as well as in cross-resistance of *M. tuberculosis* to bedaquiline and clofazimine [17]. We have also identified a possible operator sequence in the 5′-untranslated region of *MSMEG_1381*, similar to the one described in [16]: 5′-AAGCGGATTGACCTTATCCACTT-3′.

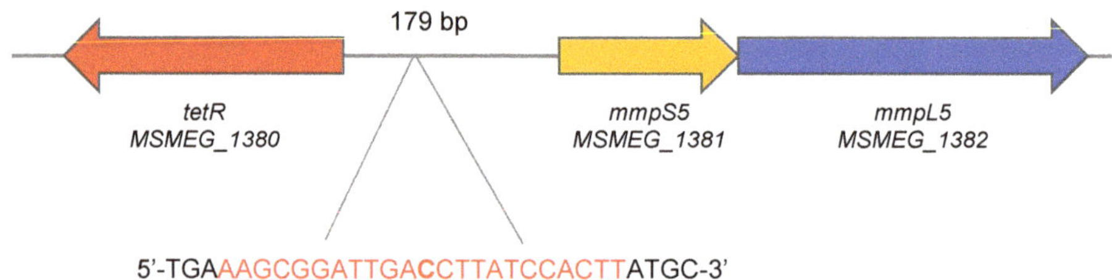

Figure 2. Schematic representation of the *mmpS5-mmpL5* operon structure in *M. smegmatis* genome. The putative operator sequence is shown in red.

To test the hypothesis that resistance to imidazo[1,2-*b*][1,2,4,5]tetrazines in *M. smegmatis* has a similar origin to the ones described for *M. tuberculosis* and *M. abscessus* [16,17], we analyzed the expression of *MSMEG_1380* gene and *mmpL5* (*MSMEG_1382*) genes in different conditions.

All the spontaneous *M. smegmatis* mutants had increased *mmpL5* expression (54.16–80.45 times) as compared to the *w.t. M. smegmatis mc2 155* strain (Figure 3A). The overexpression of the *w.t. MSMEG_1380* gene, cloned into the pMINDKm⁻ plasmid led to a 7.90-fold repression of the *mmpL5* gene expression ($p < 0.001$, Figure 3B), confirming that *MSMEG_1380* encodes the repressor of the *mmpS5-mmpL5* operon, and explaining the drug-susceptible phenotype, observed in the *MSMEG_1380* overexpressing strain. On the contrary, the expression of *MSMEG_1380* was upregulated in the mutant strains (Figure 3A), indicating that this transcriptional repressor is self-regulatory and that mutations lead to the loss of its function.

We also observed that the addition of subinhibitory concentrations of the compound **3a** upregulated the expression of *mmpL5* in a dose-dependent manner (Figure 3C).

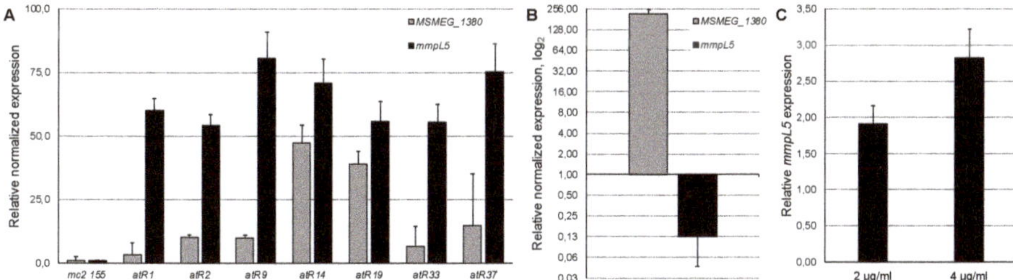

Figure 3. Relative expression levels of *mmpS5-mmpL5* operon genes in different conditions: expression levels of *MSMEG_1380* and *mmpL5* genes in spontaneous *M. smegmatis* imidazo[1,2-*b*][1,2,4,5] tetrazine-resistant mutants (**A**); expression levels of *MSMEG_1380* and *mmpL5* genes in *M. smegmatis* pMINDKm⁻:*msmeg_1380* (**B**); expression levels of the *mmpL5* gene after the addition of different concentrations of the compound **3a** (shown on the X-axis) (**C**). Error bars represent standard deviations from triplicates.

3. Discussion

Deorphaning phenotypic screening hits, that is determining their mechanism of action and/or resistance, is a key part in the early-stage anti-TB drug development [18]. Here we determine the mechanism of *M. smegmatis* imidazo[1,2-*b*][1,2,4,5]tetrazines resistance based on the previously obtained whole-genome sequencing data for 12 spontaneous mutants with cross-resistance to four compounds [9,13].

The construction of targeted mutants showed that only mutations in *MSMEG_1380* are responsible for drug resistance. In *M. smegmatis*, *MSMEG_1380* encodes a TetR-family transcriptional repressor, which controls the *mmpS5-mmpL5* operon, encoding transmembrane transporters, conserved throughout mycobacterial species. Mutations occurring in *MSMEG_1380* led to the upregulation of the *mmpS5-mmpL5* operon and increased efflux of the drug-candidates from the cells, similarly to the mechanisms described for *M. tuberculosis* and *M. abscessus* [16,17]. Overexpression of the *w.t. MSMEG_1380* led to the repression of the *mmpS5-mmpL5* operon and expectedly to an increased drug susceptibility phenotype. Interestingly, we observed a dose-dependent upregulation of the *mmpS5-mmpL5* operon upon the addition of one of the compounds, which may indicate the ability of this compound to bind to the MSMEG_1380 protein, inhibiting its affinity to the operator sequence; however, this needs to be examined in vitro in future studies.

The tested compounds showed activity as ESTPK inhibitors [9,11]; however, we have not observed any mutations in ESTPK genes. One or more ESTPKs might still be the biotargets of imidazo[1,2-*b*][1,2,4,5]tetrazines but determining them by spontaneous mutagenesis might be difficult: some of the ESTPKs may fulfill the functions of others in the situation when they might be inhibited [8], and there is a possibility that more than one mutation might be required.

The primary biological role of the MmpS5-MmpL5 system consists in siderophore transport, which is crucial for *M. tuberculosis* survival under low-iron conditions within macrophages [19]. Yet, this efflux system has also shown itself to be an important factor of drug resistance: besides the mentioned efflux-mediated resistance to thiacetazone derivatives, bedaquiline and clofazimine [16,17], it has also been reported to provide *M. tuberculosis* resistance to azoles [20]. We can expect that *M. tuberculosis* strains resistant to bedaquiline and clofazimine might also be resistant to imidazo[1,2-*b*][1,2,4,5]tetrazines; however, a 36% mismatch in the amino acid sequences of the MmpL5 proteins in *M. smegmatis* and *M. tuberculosis* may affect the drug-specificity of the transporter, and this should be additionally examined in future studies. Still, the MmpS5-MmpL5-mediated resistance mechanism needs to be considered during early stages of anti-TB drug development, and convenient in silico and in vitro test-systems for rapid analysis should be developed.

4. Materials and Methods

4.1. Bacterial Strains and Growth Conditions

M. smegmatis strains described in this study are presented in Table 1. Middlebrook 7H9 medium (Himedia, India) supplemented with OADC (Himedia, India), 0.1% Tween-80 (v/v), and 0.4% glycerol (v/v) was used as liquid medium, while the M290 Soyabean Casein Digest Agar (Himedia, India) was used as solid medium. *Escherichia coli* DH5α was used for plasmids propagation. Cultures in liquid medium were incubated in the Multitron incubator shaker (Infors HT, Switzerland) at 37 °C and 250 rpm.

4.2. Targeted *M. smegmatis* mutants' Construction

Targeted *M. smegmatis mc2 155* mutants were constructed by homologous recombination using the p2NIL/pGOAL19 suicide vector system [14]. Briefly, genes *MSMEG_0641*, *MSMEG_1380*, *MSMEG_1601*, and *MSMEG_2087* with adjacent 1-1,5 kb fragments were amplified from genomic DNA, isolated from respective mutants by phenol-chloroform/isoamyl alcohol extraction after enzymatic cell lysis [21], with Phusion High-Fidelity DNA Polymerase (Thermo Scientific, USA) using the following primers, picked with primer-BLAST [22]: pN_0641_f 5′- TTTTCTGCAGCCAACAACGATCCAGATGTCCGT-3′ and pN_0641_r 5′- TTTTAAGCTTCAATGGCGGCGTCTTCATTCTG-3′ for *MSMEG_0641*; pN_1380_f 5′- TTTTAAGCTTGTACTACTCGCTGGTGGCGTC-3′ and pN_1380_r 5′- TTTTGGATCCTGCTGCACGTG TTCGGTGTC-3′ for *MSMEG_1380*; pN_1601_f 5′- CCCATGACGGGCATCATCAACC-3′ and pN_1601_r 5′- TTTTTTAATTAACGACGATCAGCACGTCCACAC-3′ for *MSMEG_1601*; pN_2087_f 5′- TTTTAAGCTTCCAGAAGGTCACCAGCGATCTG-3′ and pN_2087_r. The amplified products were digested with respective restriction enzymes (Thermo Scientific, USA) and ligated in the p2NIL plasmid. The cassette from pGOAL19 was subsequently cloned in the obtained plasmids at the *Pac*I restriction site. The plasmids were electroporated in *M. smegmatis mc2 155* cells as described in [23] and plated on M290 plates supplemented with kanamycin (50 μg/mL), hygromycin (50 μg/mL), and X-Gal (50 μg/mL); blue single-crossover colonies were selected. Blue colonies were grown overnight in liquid 7H9 medium with ADC, and serial 10-fold dilutions were plated on M290 plates supplemented with X-Gal (50 μg/mL) and sucrose (2% w/v); white double-crossover colonies were selected and tested for Km susceptibility. Target genes were then Sanger-sequenced for a final confirmation of the mutation.

4.3. MIC Determination

MICs of the studied compounds on *M. smegmatis* were determined in liquid medium. *M. smegmatis* strains were cultured overnight in 7H9 medium, then diluted in the proportion of 1:200 in fresh medium (to approximately OD_{600} = 0.05). 196 μl of the diluted culture was poured in sterile nontreated 96-well flat-bottom culture plates (Eppendorf, Germany) and 4 μL of serial two-fold dilutions of the tested compounds in DMSO were added to the wells. The plates were incubated at 37 °C and 250 rpm for 48 h. The MIC was determined as the lowest concentration of the compound with no visible bacterial growth.

4.4. MSMEG_1380 Cloning, Expression and Drug-Susceptibility Testing

MSMEG_1380 genes from respective strains were amplified by Phusion High-Fidelity DNA Polymerase (Thermo Scientific, USA) using primers pM_1380_f 5′-GACACATATGGGAGGAAATGT TGTGAGTGCCCCCGAGACG-3′ and pM_1380_r 5′-TTTTACTAGTTCAGGTGGCGCAGGGCG-3′ picked with primer-BLAST [22] and cloned in the pMINDKm$^-$ plasmid [15], a modification of pMIND [24] lacking the kanamycin resistance gene, at the *Nde*I and *Spe*I restriction sites, to obtain the following plasmids: pMINDKm$^-$:*msmeg_1380*, pMINDKm$^-$:*msmeg_1380-19*, and pMINDKm$^-$:*msmeg_1380-33*, containing, respectively, the *w.t. msmeg_1380* gene as well as its mutant variants from strains *atR19* and *atR33*. The resulting plasmids were electroporated in *M. smegmatis mc2 155* cells as described in [23].

M. smegmatis transformants were grown in Middlebrook 7H9 broth supplemented with hygromycin (50 μg/mL) and tetracycline (10 ng/mL) to midexponential phase (OD_{600} = 1.2). Afterwards the cultures were diluted in the proportion of 1:9:10 (culture:water:M290 medium) and 5 mL were poured as the top layer on Petri dishes with agarized M290 medium. Both top- and bottom-layers were supplemented with hygromycin (50 μg/mL) and tetracycline (10 ng/mL). The plates were allowed to dry for at least 30 min, afterwards sterile paper discs with impregnated imidazo[1,2-*b*][1,2,4,5]tetrazines were plated. The plates were incubated for 2–3 days at 37 °C, until the bacterial lawn was fully grown. Growth inhibition halos were measured to the nearest 1 mm. The experiments were carried out as triplicates; the average diameter and standard deviation (SD) were calculated.

4.5. Mycobacterial RNA Isolation and Real-Time qPCR

M. smegmatis strains were grown overnight in Middlebrook 7H9 broth to midexponential phase (OD_{600} = 1.0–1.2); cells from 10 mL culture were harvested by centrifugation for 10 min at 3000× *g* and washed by 1 mL of RNAprotect Bacteria Reagent (Qiagen, USA). Total RNA was extracted by homogenization in Trizol solution (Invitrogen, USA) [25], followed by phenol (pH = 4.5)-chloroform/isoamyl alcohol (25:24:1) purification and precipitation in high salt solution (0.8 M Na citrate, 1.2 M NaCl) with isopropanol. Remaining genomic DNA was removed by DNAse I, Amplification grade (Invitrogen, USA). 50 ng of total RNA was used for cDNA synthesis by iScript Select cDNA Synthesis Kit (Bio-Rad, USA). 1 ng of cDNA was used for real-time qPCR with the qPCRmix-HS SYBR kit (Evrogen, Russia) on a CFX96 Touch machine (Bio-Rad, USA). CFX Manager V 3.1 (Bio-Rad, USA) was used to analyze the qPCR results: relative normalized expression of three biological replicates was calculated as ΔΔCq and genes *sigA* and *ftsZ* were used as reference. The following primers were picked by primer-BLAST [22] for qPCR: q1380-f 5′-CTGCTCGACGAACCATGCGAAAC-3′ and q1380-r 5′-AAGGGTCTTGAGCCGAATCTCAACG-3′ (*MSMEG_1380*), q1382-f 5′-ACCACGCAGATCATGAACAACGACT-3′ and q1382-r 5′-GAAATCGT CGAAGTCCGCCAGATGA-3′ (*MSMEG_1382*), qsigAs-sm-f 5′-CGAGCTTGTTGATCACCTCGAC CAT-3′ and qsigAs-sm-r 5′-CTCGACCTCATCCAGGAAGGCAAC-3′ (*sigA*), qftsZs-sm-f 5′-AG CAGCTCCTCGATGTCGTCCTT-3′ and qftsZs-sm-r 5′-GCCTGAAGGGCGTCGAGTTCAT-3′ (*ftsZ*).

Author Contributions: Conceptualization, D.A.M.; formal analysis, D.A.M., K.V.S., A.A.V.; investigation, D.A.M., K.V.S., A.A.V.; resources, D.A.M., K.V.S., V.N.D.; writing—original draft preparation, D.A.M.; writing review & editing, D.A.M., K.V.S., A.A.V., V.N.D.; visualization, D.A.M.; supervision, V.N.D.; project administration, D.A.M.; funding acquisition, D.A.M. All authors have read and agreed to the published version of the manuscript.

Funding: This research was funded by the Russian Science Foundation (RSF), grant number 17-75-20060.

Acknowledgments: We would like to thank Acad. V.N. Charushin and G.L. Rusinov of the Laboratory of Heterocyclic Compounds, Postovsky Institute of Organic Synthesis, Ural Branch of RAS, for generously providing compounds for this study.

Conflicts of Interest: The authors declare no conflict of interest.

References

1. *World Health Organization Global Tuberculosis Report 2019*; WHO: Geneva, Switzerland, 2019; pp. 1–297.
2. Gandhi, N.R.; Nunn, P.; Dheda, K.; Schaaf, H.S.; Zignol, M.; Van Soolingen, D.; Jensen, P.; Bayona, J. Multidrug-resistant and extensively drug-resistant tuberculosis: A threat to global control of tuberculosis. *Lancet* **2010**, *375*, 1830–1843. [CrossRef]
3. Caminero, J.A.; Sotgiu, G.; Zumla, A.; Migliori, G.B. Best drug treatment for multidrug-resistant and extensively drug-resistant tuberculosis. *Lancet Infect Dis.* **2010**, *10*, 621–629. [CrossRef]
4. Klopper, M.; Warren, R.M.; Hayes, C.; Gey van Pittius, N.C.; Streicher, E.M.; Müller, B.; Sirgel, F.A.; Chabula-Nxiweni, M.; Hoosain, E.; Coetzee, G.; et al. Emergence and Spread of Extensively and Totally Drug-Resistant Tuberculosis, South Africa. *Emerg. Infect. Dis.* **2013**, *19*, 449–455. [CrossRef] [PubMed]
5. Velayati, A.A.; Farnia, P.; Farahbod, A.M. Overview of drug-resistant tuberculosis worldwide. *Int. J. Mycobacteriol.* **2016**, *5*, S161. [CrossRef] [PubMed]

6. Muñoz-Torrico, M.; Duarte, R.; Dalcolmo, M.; D'Ambrosio, L.; Migliori, G.B. New drugs and perspectives for new anti-tuberculosis regimens. *Rev. Port. De Pneumol.* **2018**, *24*, 86–98.
7. Danilenko, V.N.; Osolodkin, D.I.; Lakatosh, S.A.; Preobrazhenskaya, M.N.; Shtil, A.A. Bacterial eukaryotic type serine-threonine protein kinases: From structural biology to targeted anti-infective drug design. *Curr. Top. Med. Chem* **2011**, *11*, 1352–1369. [CrossRef] [PubMed]
8. Prisic, S.; Husson, R.N. *Mycobacterium tuberculosis* Serine/Threonine Protein Kinases. *Microbiol. Spectr.* **2014**, *2*, 1–26. [CrossRef] [PubMed]
9. Maslov, D.A.; Korotina, A.V.; Shur, K.V.; Vatlin, A.A.; Bekker, O.B.; Tolshchina, S.G.; Ishmetova, R.I.; Ignatenko, N.K.; Rusinov, G.L.; Charushin, V.N.; et al. Synthesis and antimycobacterial activity of imidazo[1,2-*b*][1,2,4,5]tetrazines. *Eur. J. Med. Chem.* **2019**, *178*, 39–47. [CrossRef] [PubMed]
10. Bekker, O.B.; Danilenko, V.N.; Ishmetova, R.I.; Maslov, D.A.; Rusinov, G.L.; Tolshchina, S.G.; Charushin, V.N. Substituted azolo[1,2,4,5]tetrazines-inhibitors of antibacterial serine-threonine protein kinases. NPO SRC "BIOAN", Moscow, Russia. Patent RU 2462466, 27 September 2012.
11. Bekker, O.B.; Danilenko, V.N.; Maslov, D.A. Test system of *Mycobacterium smegmatis aphVIII+* for screening of inhibitors of serine-threonine protein kinases of eukaryotic type. NPO SRC "BIOAN", Moscow, Russia. Patent RU 2566998, 27 October 2015.
12. Maslov, D.A.; Bekker, O.B.; Shur, K.V.; Vatlin, A.A.; Korotina, A.V.; Danilenko, V.N. Whole-genome sequencing and comparative genomic analysis of *Mycobacterium smegmatis* mutants resistant to imidazo[1,2-*b*][1,2,4,5]tetrazines, antituberculosis drug candidates. *BRSMU* **2018**, 19–22. [CrossRef]
13. Vatlin, A.A.; Shur, K.V.; Danilenko, V.N.; Maslov, D.A. Draft Genome Sequences of 12 *Mycolicibacterium smegmatis* Strains Resistant to Imidazo[1,2-*b*][1,2,4,5]Tetrazines. *Microbiol. Resour. Announc.* **2019**, *8*, e00263-19. [CrossRef] [PubMed]
14. Parish, T.; Stoker, N.G. Use of a flexible cassette method to generate a double unmarked *Mycobacterium tuberculosis* tlyA plcABC mutant by gene replacement. *Microbiology* **2000**, *146*, 1969–1975. [CrossRef] [PubMed]
15. Maslov, D.A.; Bekker, O.B.; Alekseeva, A.G.; Kniazeva, L.M.; Mavletova, D.A.; Afanasyev, I.I.; Vasilevich, N.I.; Danilenko, V.N. Aminopyridine- and aminopyrimidine-based serine/threonine protein kinase inhibitors are drug candidates for treating drug-resistant tuberculosis. *BRSMU* **2017**, 38–43. [CrossRef]
16. Richard, M.; Gutiérrez, A.V.; Viljoen, A.J.; Ghigo, E.; Blaise, M.; Kremer, L. Mechanistic and Structural Insights Into the Unique TetR-Dependent Regulation of a Drug Efflux Pump in *Mycobacterium abscessus*. *Front. Microbio.* **2018**, *9*, 649. [CrossRef] [PubMed]
17. Andries, K.; Villellas, C.; Coeck, N.; Thys, K.; Gevers, T.; Vranckx, L.; Lounis, N.; de Jong, B.C.; Koul, A. Acquired resistance of *Mycobacterium tuberculosis* to bedaquiline. *PLoS ONE* **2014**, *9*, e102135. [CrossRef] [PubMed]
18. Cooper, C.B. Development of *Mycobacterium tuberculosis* Whole Cell Screening Hits as Potential Antituberculosis Agents. *J. Med. Chem.* **2013**, *56*, 7755–7760. [CrossRef] [PubMed]
19. Sandhu, P.; Akhter, Y. Siderophore transport by MmpL5-MmpS5 protein complex in *Mycobacterium tuberculosis*. *J. Inorg. Biochem.* **2017**, *170*, 75–84. [CrossRef] [PubMed]
20. Milano, A.; Pasca, M.R.; Provvedi, R.; Lucarelli, A.P.; Manina, G.; de Jesus Lopes Ribeiro, A.L.; Manganelli, R.; Riccardi, G. Azole resistance in *Mycobacterium tuberculosis* is mediated by the MmpS5-MmpL5 efflux system. *Tuberculosis* **2009**, *89*, 84–90. [CrossRef] [PubMed]
21. Belisle, J.T.; Mahaffey, S.B.; Hill, P.J. Isolation of Mycobacterium Species Genomic DNA. In *Mycobacteria Protocols*; Methods in Molecular Biology; Humana Press: Totowa, NJ, USA, 2010; Volume 465, pp. 1–12.
22. Ye, J.; Coulouris, G.; Zaretskaya, I.; Cutcutache, I.; Rozen, S.; Madden, T.L. Primer-BLAST: A tool to design target-specific primers for polymerase chain reaction. *BMC Bioinform.* **2012**, *13*, 134. [CrossRef] [PubMed]
23. Goude, R.; Parish, T. Electroporation of Mycobacteria. In *Mycobacteria Protocols*, 2nd ed.; Methods in Molecular Biology; Humana Press: Totowa, NJ, USA, 2010; Volume 465, pp. 203–215.
24. Blokpoel, M.C.J. Tetracycline-inducible gene regulation in mycobacteria. *Nucleic Acids Res.* **2005**, *33*, e22. [CrossRef] [PubMed]
25. Rustad, T.R.; Roberts, D.M.; Liao, R.P.; Sherman, D.R. Isolation of Mycobacterial RNA. In *Mycobacteria Protocols*; Methods in Molecular Biology; Humana Press: Totowa, NJ, USA, 2010; Volume 465, pp. 13–22.

Article

Metabolic Changes of *Mycobacterium tuberculosis* during the Anti-Tuberculosis Therapy

Julia Bespyatykh [1,*], Egor Shitikov [1,*], Dmitry Bespiatykh [1] Andrei Guliaev [1], Ksenia Klimina [1], Vladimir Veselovsky [1], Georgij Arapidi [1,2], Marine Dogonadze [3], Viacheslav Zhuravlev [3], Elena Ilina [1] and Vadim Govorun [1]

1 Federal Research and Clinical Centre of Physical-Chemical Medicine, 119435 Moscow, Russia; d.bespiatykh@gmail.com (D.B.); andrewgull87@gmail.com (A.G.); ppp843@yandex.ru (K.K.); djdf26@gmail.com (V.V.); arapidi@gmail.com (G.A.); ilinaen@gmail.com (E.I.); govorun@hotmail.ru (V.G.)

2 Moscow Institute of Physics and Technology (State University), 141701 Dolgoprudny, Russia

3 Research Institute of Phtisiopulmonology, 191036 St. Petersburg, Russia; marine-md@mail.ru (M.D.); jouravlev-slava@mail.ru (V.Z.)

* Correspondence: JuliaBespyatykh@gmail.com (J.B.); eshitikov@mail.ru (E.S.)

Received: 24 January 2020; Accepted: 15 February 2020; Published: 18 February 2020

Abstract: Tuberculosis, caused by *Mycobacterium tuberculosis* complex bacteria, remains one of the most pressing health problems. Despite the general trend towards reduction of the disease incidence rate, the situation remains extremely tense due to the distribution of the resistant forms. Most often, these strains emerge through the intra-host microevolution of the pathogen during treatment failure. In the present study, the focus was on three serial clinical isolates of *Mycobacterium tuberculosis* Beijing B0/W148 cluster from one patient with pulmonary tuberculosis, to evaluate their changes in metabolism during anti-tuberculosis therapy. Using whole genome sequencing (WGS), 9 polymorphisms were determined, which occurred in a stepwise or transient manner during treatment and were linked to the resistance (GyrA D94A; *inhA* t-8a) or virulence. The effect of the *inhA* t-8a mutation was confirmed on both proteomic and transcriptomic levels. Additionally, the amount of RpsL protein, which is a target of anti-tuberculosis drugs, was reduced. At the systemic level, profound changes in metabolism, linked to the evolution of the pathogen in the host and the effects of therapy, were documented. An overabundance of the FAS-II system proteins (HtdX, HtdY) and expression changes in the virulence factors have been observed at the RNA and protein levels.

Keywords: omics analysis; TB treatment; system analysis; Beijing B0/W148

1. Introduction

Mycobacterium tuberculosis bacteria are among the deadliest microbial pathogens in the world. Over 10 million new cases and 1.3 million lethal cases are registered annually [1]. Despite the reduction in the total tuberculosis (TB) incidence rate, the situation remains extremely tense due to the distribution of resistant forms. Among them, the multidrug-resistant- (MDR, resistance to at least isoniazid (INH) and rifampicin (RIF)) and extensively drug-resistant- (XDR, MDR phenotype with additional resistance to any fluoroquinolone (FQ) and at least one of the second-line injectable drugs capreomycin (CAP), kanamycin (KAN) or amikacin (AM)) tuberculosis pose a serious challenge for global TB control and makes successful treatment difficult or impossible [2].

Directly observed treatment (DOT) and DOT plus are the current standards of treatment strategies for the drug-susceptible and resistant forms of tuberculosis, respectively [3]. Although these protocols show high efficiency, the emergence and fixation of resistance to new antibiotics have been observed all over the world [4–6]. There are many reasons for this and not all of them have been fully elucidated, but it is worth mentioning that pathogen resistance develops in the patient, as the bacterium is a human obligate parasite. Treatment-driven positive selection leads to: the emergence of resistant clones within a population, their growth in a heteroresistant state and further ultimate fixation of a drug-resistant variant [7].

In recent years, multi-omics technologies have been increasingly used to understand the processes taking place in bacterial cells during the formation of resistance. For instance, the whole genome sequencing allows not only to undoubtedly distinguish mixed and unmixed infections in the patient, but also to define heterogeneity within one population [8–10]. The latter is often associated with the acquisition of certain drug resistance-conferring mutations, which occur independently in multiple clones and have different fitness costs, and, as a consequence, different probabilities of fixation. Apart from this, novel genetic variants may be fixed in the population by the mechanism of hitchhiking or independently [10]. A comprehensive review of these mutations has been recently published and demonstrated that most of these variants occurred in the genes, associated with the compensatory mechanisms, virulence, or relapse and survival, but their consequences remain unclear [11].

At the same time, commonly revealed mutations do not reflect all changes occurring in the cell [12]. Recently, a total of 139 genes were shown to be differentially expressed among serial isolates, while only 15 mutations were different between the strains [10]. In another work, the abundance of 46 proteins differed between resistant and sensitive strains. [13]. The *in vitro* experiments studying the effect of anti-tuberculosis drugs on bacteria have shown changes in the abundance of proteins related to amino acid, carbohydrate, and nucleotide metabolism [14].

In the present study, for the first time the author provided multi-omics analysis of three consecutive *M. tuberculosis* isolates from the same patient, to evaluate the changes in bacterial metabolism during anti-tuberculosis therapy. Based on whole genome sequencing (WGS) data, a stepwise accumulation of polymorphisms, explaining the occurrence of phenotypic resistance to fluoroquinolones and high concentrations of isoniazid, was determined. Moreover, at the proteomic and transcriptomic levels, changes were found in the loci associated with drug-resistance and virulence that only partly can be explained by mutations in the corresponding gene regions.

2. Results and Discussion

2.1. Clinical Isolates Characteristic

Three isolates from the patient with pulmonary tuberculosis have been subjected to multi-omics analysis. The first sample, RUS_B0, was isolated in 2008 before receiving TB chemotherapy. The second one, isolate 3955, was obtained one year after the treatment with high-doses of isoniazid, para-aminosalicylic acid (PASK), capreomycin, cycloserine, levofloxacin, pyrazinamide, azithromycin, clarithromycin, and amoxicillin with clavulanic acid. The 2093 sample was isolated after three years of treatment (surgery and chemotherapy according to the individual regimen).

Drug-susceptibility testing (DST) analysis has shown that the RUS_B0 isolate was MDR, while both 3955 and 2093 isolates were XDR. Phenotypic resistance to FQ evolved in the 3955 isolate after a year of treatment and was retained in the isolate 2093. Additionally, the 2093 isolate showed resistance to higher doses of isoniazid (10 mg/L).

Based on genotyping analysis, strains under study belonged to the Beijing B0/W148 cluster. Both IS*6110* RFLP and 24-loci MIRU-VNTR typing have not revealed any changes in 3955 and 2093 strains relative to RUS_B0.

2.2. Intra-host Genome Evolution of *M. tuberculosis* Serial Isolates

Three isolates of *M. tuberculosis* from a single patient, collected during different time points of treatment, were sequenced at a median depth of 383x coverage. Allele frequency of > 25% was taken as a threshold for sub-populations differentiation. In comparison to H37Rv, nine single nucleotide polymorphisms (SNPs) were variable among selected serial isolates, and seven of them were fixed in the last sample (Figure 1, Table S1).

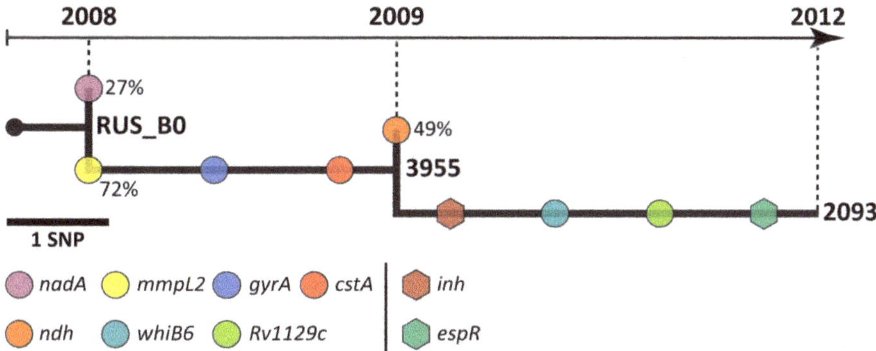

Figure 1. *M. tuberculosis* genome evolution within the patient. Sample collection time points are indicated by the numbers above the arrow. Circles and hexagons represent mutations in genes and promoter regions, respectively. Variant allele frequency is shown by the numbers beside circles.

The initial isolate RUS_B0 consisted of two populations carrying specific polymorphisms. The dominant population (72% of reads) carried synonymous SNP in the *mmpL2* gene (pos. 598098c-t; I300I), while the minor population had a substitution P287S in the NadA protein (pos. 1795614c-t).

On the second time point (12 weeks, isolate 3955), the predominant population has been ultimately fixed (all reads had a mutation in *mmpl2*) in the sample, as well as additional polymorphisms appeared: 3427440g-t (P66P) and 7582a-c (D94A) in the *cstA* and *gyrA* gene, respectively. As in the first case, the isolate 3955 consisted of two populations. One of them (49% of reads) had a 2102714g-a mutation in the *ndh* gene, leading to the T110I substitution. Of note, another substitution in this codon, T110A, was previously described as associated with isoniazid resistance, but it was found only in addition to the *ahpC* mutation [15]. Although the presence of this mutation did not affect the isoniazid resistance of the 3955 isolate, it was assumed that it was caused by the high-doses of INH used during the treatment.

It was also observed that the sub-population with a mutation in the *ndh* gene has not been fixed in the patient and in the third time point, isolate 2093, all reads corresponded to the reference. Nevertheless, a novel mutation (*inhA* t-8a), associated with resistance to INH, has emerged in this isolate. In addition to the drug resistance-associated mutations, isolate 2093 carried SNPs in *Rv1129c* (1253465c-t; S357N) and *whiB6* (4338344a-g; W60R) genes, as well as in the promoter region of the *espR* gene (c-90t).

2.3. Analysis of Drug Resistance Genes on Multi-omics Level

To evaluate the changes in bacterial metabolism during anti-tuberculosis therapy, multi-omics analysis of the serial strains was performed. The transcriptomic analysis resulted in the identification and quantification of 3889, 3894, and 3855 transcripts (TPM $\geq$ 1) in the RUS_B0, 3955, and 2093 strains, respectively (Table S2). According to the proteome analysis, 1941, 1728, and 1989 proteins were identified for the RUS_B0, 3955, and 2093 strains, respectively (Table S3). For quantitative proteomic analysis, 1358 proteins identified in all three strains were used. Over- and underrepresented proteins for each pair of strains are presented in Table S4.

According to DST results, all strains had the same following drug resistance-associated mutations: STR (RpsL: K43R), INH (KatG: S315T), RIF (RpoB: S450L), EMB (EmbB: Q497R), ETH (*ethA*: 110delA), KAN, CAP, AM (*rrs*: a1401g), PZA (*pncA*: t-11c) (Table 1). On the transcriptomic and proteomic levels, differences in the *rpoB*, *pncA*, *embB* genes and related proteins, as well as in the *rrs* gene, were not observed between the strains. Proteomic analysis revealed that in 2093 samples, abundance of RpsL and seven other ribosome proteins (Rv0055, Rv0714, Rv0720, Rv0722, Rv0979A, Rv1298, and Rv2412) was lower compared to the parental strains. This finding may be related to the fact that ribosomal proteins can be targeted by anti-tuberculosis drugs [16], and, particularly, by pyrazinamide, which was used for the treatment.

Table 1. Results of phenotypic and genetic drug susceptibility for serial strains.

Drug* (Resistance Associated Gene/Protein)	Clinical Isolates		
	RUS_B0	3955	2093
STR (Rpsl)	R (K43R)	R (K43R)	R (K43R)
INH (KatG) (*inhA*)	R (S315T)	R (S315T)	R (S315T) R+** (t-8a)
RIF (RpoB)	R (S450L)	R (S450L)	R (S450L)
EMB (EmbB)	R (Q497R)	R (Q497R)	R (Q497R)
ETH (*ethA*)	R (110delA)	R (110delA)	R (110delA)
FQ (GyrA)	S	R (D94A)	R (D94A)
KAN, CAP, AM (*rrs*)	R (a1401g)	R (a1401g)	R (a1401g)
PZA (*pncA*)	R (t-11c)	R (t-11c)	R (t-11c)

*STR - streptomycin, INH - isoniazid, RIF - rifampicin, EMB - ethambutol, ETH - ethionamide, FQ - fluoroquinolone, KAN - kanamycin, CAP - capreomycin, AM- amikacin, PZA - pyrazinamide; R - resistant, S - susceptible. **high-dose resistance (MIC=10mg/L).

For the ETH resistance-associated gene *ethA* carrying a frameshift mutation in the coding sequence, a reduction in gene expression during the treatment was found. Previously, the author has shown an increased level of *ethA* transcripts in the RUS_B0 (almost eight times) compared to the H37Rv. Its open reading frame contains numerous frame-shifts and stop codons precluding protein synthesis. The author hypothesized that MymA (Rv3083) can partly substitute the function of the inactivated gene [17]. In the present study, an up-regulation of *mymA* gene in 3955 strain compared to RUS_B0 was detected.

Starting from the second time point, the studied isolate 3955 acquired a D94A substitution in GyrA, which was also found in the 2093 isolate (Figure 1, Table 1). This mutation linked with resistance to FQ and, thus, correlates with the DST results. According to the RNA-seq and proteomics data, the levels of the *gyrA* transcript and correspondent protein were the same in all the strains. However, another gyrase subunit, GyrB, was overrepresented in the 2093 isolate.

The isoniazid-resistant strains carrying the S315T substitution in the KatG have acquired an additional mutation in the promoter region of the *inhA* gene (isolate 2093) (Figure 1, Table 1). Recently, it was shown that combination of KatG S315T and *inhA* c-15t is associated with high-level resistance (MIC $\geq$ 19.2 mg/L) to INH [18]. According to our study, the isolate 2093, harboring KatG S315T and *inhA* t-8a, was also resistant to the high drug concentrations (MIC=10 mg/L). In turn, the whole transcriptome analysis of the *mabA* – *inhA* operon revealed an increased level (7-fold) of corresponding gene transcripts in the 2093 isolate compared to the RUS_B0. It is known that INH resistance is associated with hyperproduction of the target InhA protein [19]. However, at the proteomic level, only an overabundance of MabA protein was found (changes in InhA were not detected). As mentioned above, the 2093 strain was resistant to higher concentrations of isoniazid than the RUS_B0 and 3955. InhA and MabA are both functionally and structurally related. Thus, MabA activity is also inhibited by isoniazid [20]. All of this may be evidence of high resistance of the bacterial cells.

Previously, significant differences in the levels of the *katG* and *hspX* transcripts in the INH-resistant and susceptible strains have been shown [13]. According to our data, the level of *katG* transcript was the same, while the expression of the *hspX* was higher in the 2093 and 3955 strains, compared to the RUS_B0. The abundance of KatG protein was the same after one year of therapy, but increased in the 2093 strain, which is consistent with the data, previously obtained for the INH-resistant Beijing isolates [13]. Regarding the HspX protein, its increased representation in the 3955 strain compared to RUS_B0, which went down in the strain 2093, was observed. Previously, it was proposed that HspX can be used as a marker of the disease, including its latent form [21,22]. A reduced representation of this protein in the resistant strain casts doubts on its usefulness.

2.4. Non-Specific Bacterial Response to Anti-tuberculosis Therapy

A number of studies have shown that mutations, acquired during treatment, do not reflect all metabolic changes occurring in the cell [10,12]. In our study, an increased level of the Rv2971 protein in the 2093 strain compared to the RUS_B0 and 3955 was determined. This protein is an oxidoreductase that is necessary for the growth of mycobacteria and has a potential role in detoxification of toxic metabolites. Previous studies have shown that this enzyme was differentially represented between INHr and INHs strains [23,24] and for STR [25]. In the present study, higher transcript levels of *Rv2971* in the 2093, compared to other strains, were detected. Taken together, these may reflect the strain's resistance to high-doses of isoniazid.

Previously, the author documented an increased abundance of proteins involved in lipid metabolism, in particular, those responsible for the biosynthesis of long-chain fatty acids, in strains recovered after anti-TB treatment [26]. Another study has identified differences in the abundance of Rv0241 (HtdX) and Rv3389 (HtdY) that belongs to the FAS-II system, and Rv0242c (FabG4) related to FAS-I in the INH-resistant and susceptible strains [13]. Here, the increased representation of both FAS-II system proteins in the 2093 strain, as compared to the RUS_B0, is demonstrated. However, no changes in the abundance of the FabG4 fatty acid synthase were detected. Increased representation of the Rv0469 (UmaA) protein, which is responsible for the synthesis of mycolic acids, may facilitate the formation of the cell membrane. This can prevent penetration of the anti-TB drugs, lowering their intracellular concentration and, thus, establishing favorable conditions for the survival of the pathogen and subsequent development of drug resistance. It has been frequently suggested that enzymes of the lipid metabolism play an important role in the positive selection of mycobacteria [11], and fatty acids can be considered as virulence factors of these pathogens [17,26].

2.5. Variability in the Virulence Factors Representation

It is well known that the DosR system is involved in the response of mycobacteria to the action of the host cell immune system and to the changing environmental conditions [27,28]. The author has previously shown that under normal growth conditions the DosR regulon proteins were poorly represented in the Beijing B0/W148 cluster strains compared to the H37Rv. This is correlated with an increased abundance of the same proteins and corresponding transcripts in cluster strains under stress conditions [17]. In the present study, a statistically significant increase in the transcription levels of the DosR regulon genes on the background of anti-tuberculosis therapy was observed (Figure 2). The level of all DosR regulon genes has increased in the 3955 and 2093 compared to the RUS_B0 strain, except for the Rv1734c and Rv1812c genes, of which transcription levels were lower in the 3955 strain.

A mutation in the promoter region of *espR* was only detected in the 2093 strain. It has led to a decrease in the protein abundance and transcript level (albeit, not statistically significant). The virulence phenotype of bacteria is often implemented through protein hyperproduction *in vivo* [29], which is mediated by some intermediate activators. This also explains the low representation of the EspR and EspC proteins in the analyzed strain.

In the present study, a low abundance of the PtpA protein after the anti-TB therapy in the 2093 strain was observed. According to the transcriptome analysis, expression of the *ptpA* was higher in the 2093 strain. Concordantly, representation of the bifunctional protein Rv2228c, which is associated with PtpA, was higher. At the same time, the level of the *Rv2228c* transcript was higher in the 3955 strain, compared to the RUS_B0. It is known that PtpA (Rv2234) and PtpB (Rv0153c) are secreted proteins, which play an important role in the pathogen interaction with the host cell [30] and have orthologs among other pathogenic pro- and eukaryotes [31]. Rv2228c is presumably involved in bacterial replication, as it is only authenticated RNase HI in *M. tuberculosis*.

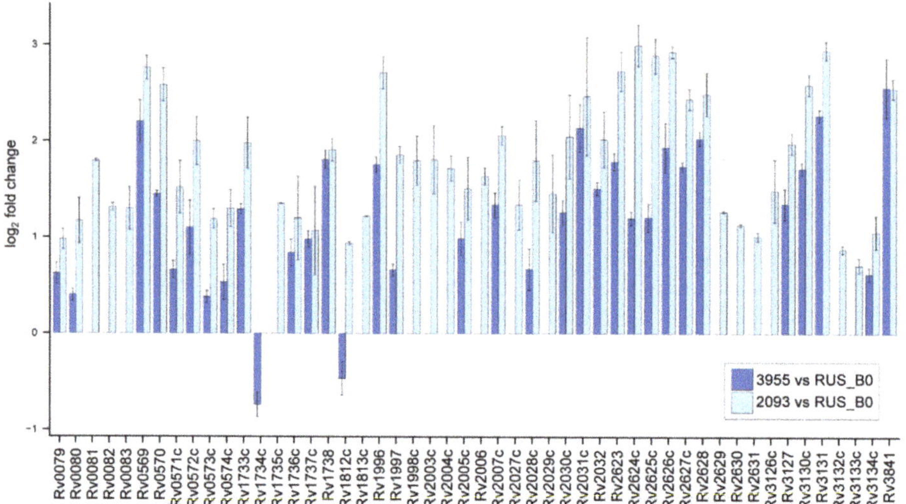

Figure 2. DosR regulon expression. Bars indicate fold changes in gene expression in 3955 (blue bars) and 2093 (light blue bars) strains compared to the RUS_B0 strain. Error bars depict standard error of the mean from three independent experiments.

3. Materials and Methods

3.1. *Mycobacterium Tuberculosis Strains and Growth Conditions*

Three strains (RUS_B0, 3955, 2093) of *Mycobacterium tuberculosis* Beijing B0/W148 cluster from the same patient were used. Strains were grown in liquid Middlebrook 7H9 medium with Oleic Albumin Dextrose Catalase (OADC) supplement (Becton Dickinson, Franklin Lakes, USA). The cultures (90 mL) were grown in three biological replicates per strain in a cell culture flask (T175, Eppendorf, Germany) in a horizontal position. *M. tuberculosis* strains were incubated at 37°C for 12 days with constant shaking (5 rpm) until OD_{600} was reached to 0.4 [32]. The bacterial suspension from each flask was immediately aliquoted in a sterile 50 mL (for transcriptome and proteome analysis) and 15 mL (for WGS) tubes at room temperature (RT).

For WGS, an aliquot (10 mL) was centrifuged at 3200g for 10 min (RT) and the pellet was stored at -20 °C. Samples (40 mL) for transcriptomic analysis were centrifuged at 3200g for 10 min (RT) and cells pellet was frozen in liquid nitrogen and stored at −80 °C until use. For proteomic analysis, cells (40 mL of culture) were harvested by centrifugation at 3500g, 4°C for 5 min and washed with Tris-HCl and 2% Triton-X100 (pH 7.5–8). Cells were precipitated by centrifugation at 4500g, 4°C for 15 min. The pellet was frozen in liquid nitrogen and stored at −80 °C until required.

The susceptibility testing was done using the BACTEC MGIT 960 Culture system (Becton Dickinson) following the manufacturer's protocol.

3.2. *Genomic Analysis*

Genomic DNA was isolated from the strains using standard extraction methods [33]. To verify that the strains belong to a Beijing B0/W148 cluster, the PCR assay was performed as described previously [34]. Spoligotyping, IS*6110*-RFLP and 24-MIRU-VNTR typing were performed as described in [35], [33], and [36], respectively.

Whole genome sequencing of the strains was performed on Illumina HiSeq 2500 Sequencing Platform according to the manufacturer's instructions. Raw data were deposited in the NCBI Sequence Read Archive under accession number PRJNA421323. The circular genome of RUS_B0 strains was assembled and deposited in the NCBI earlier (CP030093.1) [37]. WGS reads were aligned to the *M. tuberculosis* H37Rv (NC_000962.3) and RUS_B0 (CP030093.1) genome sequences using Bowtie 2 [38]. SAMtools (v.0.1.18) [39], FreeBayes (v.1.1.0) [40] and Snippy (v.4.3.6) [41] were used for variant calling [39,42]. For FreeBayes and Snippy SNPs with a minimum mapping quality of 20, minimum coverage of 10 and alternate fraction of 0.9 were taken.

A comprehensive list of drug-resistance mutations to first- and second-line drugs were used to determine genetically resistant strains [43]. The identification of IS*6110* integration sites was carried out using ISMapper pipeline [44].

3.3. Transcriptomic Analysis

3.3.1. RNA Extraction

Total RNA was isolated from all *M. tuberculosis* cultures as previously described [32]. In brief, bacteria were re-suspended in 1 mL Trizol (Invitrogen) and added to a 2-mL Lysing Matrix B (MP Bio, USA). Cells were disrupted by bead-beating twice for 1 min with a 2-min interval on ice. The suspension was then transferred to a new tube, where chloroform extraction was performed. RNA was precipitated by adding 0.7 volume of isopropanol and washed with 70% ethanol, air-dried and re-suspended in 100 μL DEPC-treated water.

DNase treatment was carried out with TURBO DNA-free kit (Thermo Fisher Scientific) in volumes of 100 μL and further with the RNase-Free DNase Set (Qiagen, Germany) according to the manufacturer's protocol. RNA cleanup was performed with the RNeasy Mini Kit (Qiagen) according to the RNA Cleanup protocol and stored at −70°C until further use. DNA contamination was evaluated by PCR, using primers for amplification of the IS*6110* fragment (PolyTub, Lytech, Russia). The concentration and quality of the total extracted RNA were checked by using the Quant-it RiboGreen RNA assay (Thermo Fisher Scientific) and the RNA 6000 pico chip (Agilent Technologies), respectively.

3.3.2. RNA-seq and Analysis

Total RNA (1 - 2.5 μg) was used for library preparation. Ribosomal RNA was removed from the total RNA and libraries were prepared using the ScriptSeq Complete kit (Epicentre/Illumina, Madison, USA), according to the manufacturer's protocol. Subsequently, RNA cleanup was performed with the Agencourt RNA Clean XP kit (Beckman Coulter, Brea, USA). The library underwent a final cleanup using the Agencourt AMPure XP system (Beckman Coulter) after which the libraries' size distribution and quality were assessed using a high sensitivity DNA chip (Agilent Technologies). Libraries were subsequently quantified by Quant-iT DNA Assay Kit, High Sensitivity (Thermo Fisher Scientific). Finally, equimolar quantities of all libraries (12 pM) were sequenced by a high throughput run on the Illumina HiSeq using 2×125 bp paired-end reads and a 5% Phix spike-in control. Before loading the cBot system, the libraries were incubated at 98°C for 2 minutes and then cooled on ice to improve the hybridization of the GC-rich sequences. In total, 114 and 110 million paired reads were obtained corresponding to 14 and 13 billion nucleotide bases for H37Rv and RUS_B0, respectively. The dataset of RNA-Seq analysis was deposited to the NCBI with the project name PRJNA421323.

Adaptors were trimmed with the Trimmomatic v0.33 tool [45]. Quality control on raw reads was carried out with FASTQC v0.11.5 [46]. The Kallisto v0.46.0 [47] software was used for the reads mapping and abundance estimation. Differential expression analysis was performed using edgeR v3.26.8 [48] package, integrated in the Degust v4.1.1 [49] web-tool. Only genes with count per million (CPM) ≥ 1 were analyzed further. Genes were filtered based on false discovery rate cutoff (FDR) ≤ 0.05 and minimum expression fold change (FC) ≥ 2. The plots were generated using the ggplot2 package in R.

3.4. Proteomic Analysis

3.4.1. Protein Extraction and Trypsin Digestion

The bacterial pellet was lysed and protein was extracted as described previously [17]. Protein concentration was measured by the Bradford method [50] using the Bradford Protein Assay Kit (Bio-Rad, Hercules, USA).

Proteolytic in-gel digestion was performed in three biological and two technical replicates: 1) sample, fractionated into 6 parts; 2) total load sample as described previously [51]. Peptides were cleaned using C18 Sep-Pak columns (Waters, Milford, USA) [17].

3.4.2. LC-MS/MS Analysis

Peptides were separated with high-performance liquid chromatography (HPLC, Ultimate 3000 Nano LC System, Thermo Fisher Scientific,) in a 15-cm long C18 column with a diameter of 75 µm (Acclaim®PepMap™ RSLC, Thermo Fisher Scientific). The peptides were eluted with a gradient from 5% to 35% of buffer B (80% acetonitrile, 0.1% formic acid) over 115 min at a flow rate of 0.3 µl/min. Total run time including 15 min to reach 99% buffer B, flushing 10 min with 99% buffer B and 15 min re-equilibration to buffer A (0.1% formic acid) amounted to 65 min. Further analysis was performed with a Q Exactive HF mass spectrometer (Q ExactiveTM HF Hybrid Quadrupole-OrbitrapTM Mass spectrometer, Thermo Fisher Scientific). Mass spectra were acquired at a resolution of 60,000 (MS) and 15,000 (MS/MS) in a range of 400-1500 m/z (MS) and 200-2000 m/z (MS/MS). An isolation threshold of 67,000 was determined for precursor selection, and up to the top 25 precursors were chosen for fragmentation with high-energy collisional dissociation (HCD) at 25 V and 100 ms activation time. Precursors with a charged state of +1 were rejected and all measured precursors were excluded from measurement for 20 s. From 2 to 4 technical runs were analyzed for each sample.

3.4.3. Protein Identification and Quantitation

Raw data was captured from the mass spectrometer and converted to MGF (Mascot Generic Format) files using ProteoWizard with the following parameters: peakPicking true 2, msLevel 2, zeroSamples removeExtra [52]. For thorough protein identification, the generated peak lists were searched with the MASCOT (v 2.5.1, Matrix Science Ltd, UK) and X! Tandem (VENGEANCE, 2015.12.15, The Global Proteome Machine Organization) search engines. Database-searching parameters were as follows: tryptic hydrolysis, no more than one missed site, the precursor and fragment mass tolerance were set at 20 ppm and 50 ppm, respectively. Oxidation of methionine was set as a possible modification, carbamidomethylation of cysteine as a fixed. For X! Tandem, parameters that allowed a quick check for protein N-terminal residue acetylation, peptide N-terminal glutamine ammonia loss or peptide N-terminal glutamic acid water loss were selected. Resulting files were submitted to the Scaffold 4 software (v 4.2.1, Proteome Software, Inc, USA) for validation and further analysis. For protein identification, the proteomic databases for the RUS_B0 (RefSeq: CP030093.1) and reference strain H37Rv (RefSeq: NC_000962.3) genomes were used. Additionally, in proteome searching database peptides containing strain-specific single amino acid polymorphisms were added according to the approach described earlier [51]. The local false discovery rate scoring algorithm with standard experiment-wide protein grouping was used. A 1% FDR threshold was applied to search results from individual datasets. For all detected proteins, functional categories (TubercuList v 2.6 (http://tuberculist.epfl.ch/)) and subcellular localizations (PSORTdb v 3.0 (http://db.psort.org/)) were established.

For label-free quantitation, raw MS data files (.wiff files) were imported and processed in Progenesis LC-MS software v.4.1 (Nonlinear Dynamics, Newcastle, UK). The results of peptide quantitation were normalized using an iterative median-based normalization as implemented in the Progenesis software. Differences in the abundance of a protein between the three biological replicates of all strains were evaluated using a two-sided unpaired Student's T-test. P-values < 0.05 were considered statistically significant. Adjusted *p*-values for multiple tests (*q*-values) were generated using the Benjamini–Hochberg method [53].

4. Conclusions

This is a first report of the system omics analysis of *M. tuberculosis* Beijing B0/W148 cluster strains isolated from one patient at different stages of TB therapy. A clear correlation between genetic and phenotypic resistance to the antibiotics was demonstrated, documenting changes in the lipid metabolism, and production of virulence factors on the proteomic and transcriptomic levels. These changes imply that the anti-tuberculosis therapy play a crucial role in the regulation of bacterial biochemical pathways.

We show that integrating various approaches can advance our knowledge of the role and significance of microevolution in *M. tuberculosis* infection.

Supplementary Materials: The following are available online at http://www.mdpi.com/2076-0817/9/2/131/s1. Table S1: Variable SNPs among tree clinical isolates. Table S2: List of identified and quantified transcripts for RUS_B0, 3955 and 2093 strains. Table S3: Proteins and peptides identified in RUS_B0, 3955 and 2093 strains. Table S4: Quantitative proteomic analysis of RUS_B0, 3955 and 2093 strains.

Author Contributions: Conceptualization, J.B. and E.S.; methodology, J.B., E.S., K.K., M.D. and V.V.; software, D.B., A.G., and G.A.; formal analysis, D.B., A.G., G.A., K.K., M.D. and V.V.; investigation, J.B., E.S.; resources, V.Z., E.I. and V.G.; data curation, J.B., E.S. and V.Z.; writing – original draft preparation, J.B. and E.S.; writing – review & editing, J.B, E.S. and D.B.; supervision, E.I.; project administration, V.G.; funding acquisition, J.B. All authors have read and agreed to the published version of the manuscript.

Funding: This work was supported by Russian Foundation for Basic Research (project 18-34-00168).

Acknowledgments: We acknowledge the IBMC "Human Proteome" Core Facility for assistance with the generation of mass spectrometry data. We thank the Center for Precision Genome Editing and Genetic Technologies for Biomedicine, Federal Research and Clinical Center of Physical-Chemical Medicine of Federal Medical Biological Agency for providing the WGS platform.

Conflicts of Interest: The authors declare no conflict of interest.

References

1. WHO. *Weekly Epidemiological Record*; WHO: Geneva, Switzerland, 2020; Volume 95, pp. 1–12.
2. WHO. *Weekly Epidemiological Record No 45*; WHO: Geneva, Switzerland, 2006; Volume 81, pp. 425–432.
3. WHO. *Treatment of TB: Guidelines for National Programmes*; WHO/TB/97.220; WHO: Geneva, Switzerland, 1997.
4. Calver, A.D.; Falmer, A.A.; Murray, M.; Strauss, O.J.; Streicher, E.M.; Hanekom, M.; Liversage, T.; Masibi, M.; Van Helden, P.D.; Warren, R.M.; et al. Emergence of increased resistance and extensively drug-resistant tuberculosis despite treatment adherence, South Africa. *Emerg. Infect. Dis.* **2010**, *16*, 264–271. [CrossRef] [PubMed]
5. Dye, C. Doomsday postponed? Preventing and reversing epidemics of drug-resistant tuberculosis. *Nat. Rev. Microbiol.* **2009**, *7*, 81–87. [CrossRef] [PubMed]
6. Müller, B.; Borrell, S.; Rose, G.; Gagneux, S. The heterogeneous evolution of multidrug-resistant *Mycobacterium tuberculosis*. *Trends Genet.* **2013**, *29*, 160–169. [CrossRef] [PubMed]
7. Didelot, X.; Walker, A.S.; Peto, T.E.; Crook, D.W.; Wilson, D.J. Within-host evolution of bacterial pathogens. *Nat Rev Microbiol.* **2016**, *14*, 150–162. [CrossRef]
8. Sun, G.; Luo, T.; Yang, C.; Dong, X.; Li, J.; Zhu, Y.; Zheng, H.; Tian, W.; Wang, S.; Barry, C.E.; et al. Dynamic population changes in *Mycobacterium tuberculosis* during acquisition and fixation of drug resistance in patients. *J. Infect. Dis.* **2012**, *206*, 1724–1733. [CrossRef]
9. Merker, M.; Kohl, T.A.; Roetzer, A.; Truebe, L.; Richter, E.; Rüsch-Gerdes, S.; Fattorini, L.; Oggioni, M.R.; Cox, H.; Varaine, F.; et al. Whole genome sequencing reveals complex evolution patterns of multidrug-resistant *Mycobacterium tuberculosis* Beijing strains in patients. *PLoS ONE* **2013**, *8*, e82551. [CrossRef]
10. Eldholm, V.; Norheim, G.; von der Lippe, B.; Kinander, W.; Dahle, U.R.; Caugant, D.A.; Mannsåker, T.; Mengshoel, A.T.; Dyrhol-Riise, A.M.; Balloux, F. Evolution of extensively drug-resistant *Mycobacterium tuberculosis* from a susceptible ancestor in a single patient. *Genome Biol.* **2014**, *15*, 490. [CrossRef]

11. Ley, S.D.; de Vos, M.; Van Rie, A.; Warren, R.M. Deciphering Within-Host Microevolution of Mycobacterium tuberculosis through Whole-Genome Sequencing: The Phenotypic Impact and Way Forward. *Microbiol. Mol. Biol. Rev.* **2019**, *83*, e00062-18. [CrossRef]
12. Davies, A.P.; Billington, O.J.; McHugh, T.D.; Mitchison, D.A.; Gillespie, S.H. Comparison of phenotypic and genotypic methods for pyrazinamide susceptibility testing with *Mycobacterium tuberculosis*. *J. Clin. Microbiol.* **2000**, *38*, 3686–3688. [CrossRef]
13. María Nieto, L.R.; Mehaffy, C.; Dobos, K.M. Comparing isogenic strains of Beijing genotype *Mycobacterium tuberculosis* after acquisition of Isoniazid resistance: A proteomics approach. *Proteomics* **2016**, *16*, 1376–1380. [CrossRef]
14. Danelishvili, L.; Shulzhenko, N.; Chinison, J.J.J.; Babrak, L.; Hu, J.; Morgun, A.; Burrows, G.; Bermudeza, L.E. *Mycobacterium tuberculosis* proteome response to antituberculosis compounds reveals metabolic "escape" pathways that prolong bacterial survival. *Antimicrob. Agents Chemother.* **2017**, *61*, e00430-17. [CrossRef] [PubMed]
15. Lee, A.S.G.; Teo, A.S.M.; Wang, S.Y. Novel mutations in ndh in isoniazid-resistant Mycobacterium tuberculosis isolates. *Antimicrob. Agents Chemother.* **2001**, *45*, 2157–2159. [CrossRef] [PubMed]
16. Finken, M.; Kirschner, P.; Meier, A.; Wrede, A.; Böttger, E.C. Molecular basis of streptomycin resistance in *Mycobacterium tuberculosis*: Alterations of the ribosomal protein S12 gene and point mutations within a functional 16S ribosomal RNA pseudoknot. *Mol. Microbiol.* **1993**, *9*, 1239–1246. [CrossRef] [PubMed]
17. Bespyatykh, J.; Shitikov, E.; Butenko, I.; Altukhov, I.; Alexeev, D.; Mokrousov, I.; Dogonadze, M.; Zhuravlev, V.; Yablonsky, P.; Ilina, E.; et al. Proteome analysis of the *Mycobacterium tuberculosis* Beijing B0/W148 cluster. *Sci. Rep.* **2016**, *6*, 28985. [CrossRef]
18. Lempens, P.; Meehan, C.J.; Vandelannoote, K.; Fissette, K.; De Rijk, P.; Van Deun, A.; Rigouts, L.; De Jong, B.C. Isoniazid resistance levels of *Mycobacterium tuberculosis* can largely be predicted by high-confidence resistance-conferring mutations. *Sci. Rep.* **2018**, *8*, 3246. [CrossRef]
19. Banerjee, A.; Dubnau, E.; Quemard, A.; Balasubramanian, V.; Um, K.S.; Wilson, T.; Collins, D.; De Lisle, G.; Jacobs, W.R. inhA, a gene encoding a target for isoniazid and ethionamide in Mycobacterium tuberculosis. *Science* **1994**, *263*, 227–230. [CrossRef]
20. Ducasse-Cabanot, S.; Cohen-Gonsaud, M.; Marrakchi, H.; Nguyen, M.; Zerbib, D.; Bernadou, J.; Daffé, M.; Labesse, G.; Quémard, A.A. In Vitro Inhibition of the *Mycobacterium tuberculosis*-Ketoacyl-Acyl Carrier Protein Reductase MabA by Isoniazid. *Antimicrob. Agents Chemother.* **2004**, *48*, 242–249. [CrossRef]
21. Castro-Garza, J.; García-Jacobo, P.; Rivera-Morales, L.G.; Quinn, F.D.; Barber, J.; Karls, R.; Haas, D.; Helms, S.; Gupta, T.; Blumberg, H.; et al. Detection of anti-HspX antibodies and HspX protein in patient sera for the identification of recent latent infection by *Mycobacterium tuberculosis*. *PLoS ONE* **2017**, *12*, e0181714. [CrossRef]
22. Dhiman, A.; Haldar, S.; Mishra, S.K.; Sharma, N.; Bansal, A.; Ahmad, Y.; Kumar, A.; Sharma, T.K.; Tyagi, J.S. Generation and application of DNA aptamers against HspX for accurate diagnosis of tuberculous meningitis. *Tuberculosis* **2018**, *112*, 27–36. [CrossRef]
23. Jiang, X.; Zhang, W.; Gao, F.; Huang, Y.; Lv, C.; Wang, H. Comparison of the proteome of isoniazid-resistant and -susceptible strains of *Mycobacterium tuberculosis*. *Microb. Drug Resist.* **2006**, *12*, 231–238. [CrossRef]
24. Phong, T.Q.; Ha, D.T.T.; Volker, U.; Hammer, E. Using a Label Free Quantitative Proteomics Approach to Identify Changes in Protein Abundance in Multidrug-Resistant *Mycobacterium tuberculosis*. *Indian J. Microbiol.* **2015**, *55*, 219–230. [CrossRef] [PubMed]
25. Sharma, P.; Kumar, B.; Gupta, Y.; Singhal, N.; Katoch, V.M.; Venkatesan, K.; Bisht, D. Proteomic analysis of streptomycin resistant and sensitive clinical isolates of *Mycobacterium tuberculosis*. *Proteome Sci.* **2010**, *8*, 59. [CrossRef] [PubMed]
26. Bespyatykh, J.; Shitikov, E.; Zgoda, V.; Smolyakov, A.; Zamachaev, M.; Dogonadze, M.; Zhuravlev, V.; Ilina, E. Changes in metabolism of *Mycobacterium tuberculosis* Beijing B0/W148 cluster against the background of anti-tuberculosis therapy. *FEBS J.* **2017**, *284* (Suppl. 1), 347–347.
27. Dasgupta, N.; Kapur, V.; Singh, K.K.; Das, T.K.; Sachdeva, S.; Jyothisri, K.; Tyagi, J.S. Characterization of a two-component system, devR-devS, of Mycobacterium tuberculosis. *Tuber. Lung Dis.* **2000**, *80*, 141–159. [CrossRef] [PubMed]

28. Converse, P.J.; Karakousis, P.C.; Klinkenberg, L.G.; Kesavan, A.K.; Ly, L.H.; Allen, S.S.; Grosset, J.H.; Jain, S.K.; Lamichhane, G.; Manabe, Y.C.; et al. Role of the dosR-dosS two-component regulatory system in *Mycobacterium tuberculosis* virulence in three animal models. *Infect. Immun.* **2009**, *77*, 1230–1237. [CrossRef]
29. Cao, G.; Howard, S.T.; Zhang, P.; Hou, G.; Pang, X. Functional analysis of the EspR binding sites upstream of *espR* in *Mycobacterium tuberculosis*. *Curr. Microbiol.* **2013**, *67*, 572–579. [CrossRef]
30. Koul, A.; Choidas, A.; Treder, M.; Tyagi, A.K.; Drlica, K.; Singh, Y.; Ullrich, A. Cloning and characterization of secretory tyrosine phosphatases of *Mycobacterium tuberculosis*. *J. Bacteriol.* **2000**, *182*, 5425–5432. [CrossRef]
31. Kraeva, N.; Leštinová, T.; Ishemgulova, A.; Majerová, K.; Butenko, A.; Vaselek, S.; Bespyatykh, J.; Charyyeva, A.; Spitzová, T.; Kostygov, A.Y.; et al. *LmxM*.22.0250-Encoded Dual Specificity Protein/Lipid Phosphatase Impairs *Leishmania mexicana* Virulence In Vitro. *Pathogens* **2019**, *8*, 241. [CrossRef]
32. Benjak, A.; Sala, C.; Hartkoorn, R.C. Whole-transcriptome sequencing for high-resolution transcriptomic analysis in *Mycobacterium tuberculosis*. *Methods Mol. Biol.* **2015**, *1285*, 17–30.
33. Van Embden, J.D.; Cave, M.D.; Crawford, J.T.; Dale, J.W.; Eisenach, K.D.; Gicquel, B.; Hermans, P.; Martin, C.; McAdam, R.; Shinnick, T.M. Strain identification of *Mycobacterium tuberculosis* by DNA fingerprinting: Recommendations for a standardized methodology. *J. Clin. Microbiol.* **1993**, *31*, 406–409. [CrossRef]
34. Mokrousov, I.; Vyazovaya, A.; Zhuravlev, V.; Otten, T.; Millet, J.; Jiao, W.W.; Shen, A.D.; Rastogi, N.; Vishnevsky, B.; Narvskaya, O. Real-time PCR assay for rapid detection of epidemiologically and clinically significant *Mycobacterium tuberculosis* Beijing genotype isolates. *J. Clin. Microbiol.* **2014**, *52*, 1691–1693. [CrossRef] [PubMed]
35. Bespyatykh, J.A.; Zimenkov, D.V.; Shitikov, E.A.; Kulagina, E.V.; Lapa, S.A.; Gryadunov, D.A.; Ilina, E.N.; Govorun, V.M. Spoligotyping of *Mycobacterium tuberculosis* complex isolates using hydrogel oligonucleotide microarrays. *Infect. Genet. Evol.* **2014**, *26*, 41–46. [CrossRef] [PubMed]
36. Supply, P.; Allix, C.; Lesjean, S.; Cardoso-Oelemann, M.; Rüsch-Gerdes, S.; Willery, E.; Savine, E.; de Haas, P.; van Deutekom, H.; Roring, S.; et al. Proposal for standardization of optimized mycobacterial interspersed repetitive unit-variable-number tandem repeat typing of *Mycobacterium tuberculosis*. *J. Clin. Microbiol.* **2006**, *44*, 4498–4510. [CrossRef] [PubMed]
37. Bespyatykh, J.; Shitikov, E.; Andrei, G.; Smolyakov, A.; Klimina, K.; Veselovsky, V.; Malakhova, M.; Georgij, A.; Dogonadze, M.; Manicheva, O.; et al. System OMICs analysis of *Mycobacterium tuberculosis* Beijing B0/W148 cluster. *Sci. Rep.* **2019**, *9*, 1–11. [CrossRef] [PubMed]
38. Langmead, B.; Salzberg, S.L. Fast gapped-read alignment with Bowtie 2. *Nat. Methods* **2012**, *9*, 357–359. [CrossRef] [PubMed]
39. Ramirez-Gonzalez, R.H.; Bonnal, R.; Caccamo, M.; Maclean, D. Bio-samtools: Ruby bindings for SAMtools, a library for accessing BAM files containing high-throughput sequence alignments. *Source Code Biol. Med.* **2012**, *7*, 6. [CrossRef] [PubMed]
40. Garrison, E.; Marth, G. Haplotype-based variant detection from short-read sequencing. *arXiv* **2012**, arXiv:1207.3907.
41. Seemann, T. GitHub-Tseemann/Snippy: Rapid Haploid Variant Calling and Core Genome Alignment. 2015. Available online: https://github.com/tseemann/snippy (accessed on 23 January 2020).
42. Nurk, S.; Bankevich, A.; Antipov, D.; Gurevich, A.; Korobeynikov, A.; Lapidus, A.; Prjibelsky, A.; Pyshkin, A.; Sirotkin, A.; Sirotkin, Y.; et al. *Assembling Genomes and Mini-Metagenomes from Highly Chimeric Reads*; Springer: Berlin/Heidelberg, Germany, 2013; pp. 158–170.
43. Schleusener, V.; Köser, C.U.; Beckert, P.; Niemann, S.; Feuerriegel, S. *Mycobacterium tuberculosis* resistance prediction and lineage classification from genome sequencing: Comparison of automated analysis tools. *Sci. Rep.* **2017**, *7*, 46327. [CrossRef]
44. Hawkey, J.; Hamidian, M.; Wick, R.R.; Edwards, D.J.; Billman-Jacobe, H.; Hall, R.M.; Holt, K.E. ISMapper: Identifying transposase insertion sites in bacterial genomes from short read sequence data. *BMC Genom.* **2015**, *16*, 667. [CrossRef]
45. Bolger, A.M.; Lohse, M.; Usadel, B. Trimmomatic: A flexible trimmer for Illumina sequence data. *Bioinformatics* **2014**, *30*, 2114–2120. [CrossRef]
46. Andrews, S. FastQC: A Quality Control Tool for High Throughput Sequence Data. *Babraham Bioinform.* Available online: http://www.bioinformatics.babraham.ac.uk/projects/fastqc/ (accessed on 24 January 2020).
47. Bray, N.L.; Pimentel, H.; Melsted, P.; Pachter, L. Near-optimal probabilistic RNA-seq quantification. *Nat. Biotechnol.* **2016**, *34*, 525–527. [CrossRef] [PubMed]

48. Robinson, M.D.; Mccarthy, D.J.; Smyth, G.K. edgeR: A Bioconductor package for differential expression analysis of digital gene expression data. *Bioinformatics* **2010**, *26*, 139–140. [CrossRef] [PubMed]
49. Powell, D.R. Degust: Interactive RNA-seq Analysis. Available online: http://victorian-bioinformatics-consortium.github.io/degust/ (accessed on 20 January 2020).
50. Bradford, M.M. A rapid and sensitive method for the quantitation of microgram quantities of protein utilizing the principle of protein-dye binding. *Anal. Biochem.* **1976**, *72*, 248–254. [CrossRef]
51. Bespyatykh, J.; Smolyakov, A.; Guliaev, A.; Shitikov, E.; Arapidi, G.; Butenko, I.; Dogonadze, M.; Manicheva, O.; Ilina, E.; Zgoda, V.; et al. Proteogenomic analysis of *Mycobacterium tuberculosis* Beijing B0/W148 cluster strains. *J. Proteom.* **2019**, *192*, 18–26. [CrossRef] [PubMed]
52. Kessner, D.; Chambers, M.; Burke, R.; Agus, D.; Mallick, P. ProteoWizard: Open source software for rapid proteomics tools development. *Bioinformatics* **2008**, *24*, 2534–2536. [CrossRef] [PubMed]
53. Benjamini, Y.; Hochberg, Y. Controlling the False Discovery Rate: A Practical and Powerful Approach to Multiple Testing. *J. R. Stat.* **1995**, *57*, 289–300. [CrossRef]

Article

Genome Subtraction and Comparison for the Identification of Novel Drug Targets against *Mycobacterium avium* subsp. *hominissuis*

Reaz Uddin [1,*], Bushra Siraj [1], Muhammad Rashid [1], Ajmal Khan [2], Sobia Ahsan Halim [2] and Ahmed Al-Harrasi [2,*]

1 Dr. Panjwani Center for Molecular Medicine and Drug Research, International Center for Chemical and Biological Sciences, University of Karachi, Karachi 75270, Pakistan; bushrasiraj52@gmail.com (B.S.); m.rashid@iccs.edu (M.R.)

2 Natural and Medical Sciences Research Center, University of Nizwa, P.O. Box 33, Birkat Al Mauz, Nizwa 616, Sultanate of Oman; ajmalkhan@unizwa.edu.om (A.K.); sobia_halim@unizwa.edu.om (S.A.H.)

* Correspondence: mriazuddin@iccs.edu (R.U.); aharrasi@unizwa.edu.om (A.A.-H.); Tel.: +92-21-34824930 (R.U.); +96825446328 (A.A.-H.)

Received: 10 February 2020; Accepted: 26 April 2020; Published: 12 May 2020

Abstract: *Mycobacterium avium* complex (MAC) is a major cause of non-tuberculous pulmonary and disseminated diseases worldwide, inducing bronchiectasis, and affects HIV and immunocompromised patients. In MAC, *Mycobacterium avium* subsp. *hominissuis* is a pathogen that infects humans and mammals, and that is why it is a focus of this study. It is crucial to find essential drug targets to eradicate the infections caused by these virulent microorganisms. The application of bioinformatics and proteomics has made a significant impact on discovering unique drug targets against the deadly pathogens. One successful bioinformatics methodology is the use of in silico subtractive genomics. In this study, the aim was to identify the unique, non-host and essential protein-based drug targets of *Mycobacterium avium* subsp. *hominissuis* via in silico a subtractive genomics approach. Therefore, an in silico subtractive genomics approach was applied in which complete proteome is subtracted systematically to shortlist potential drug targets. For this, the complete dataset of proteins of *Mycobacterium avium* subsp. *hominissuis* was retrieved. The applied subtractive genomics method, which involves the homology search between the host and the pathogen to subtract the non-druggable proteins, resulted in the identification of a few prioritized potential drug targets against the three strains of *M. avium* subsp. *Hominissuis*, i.e., MAH-TH135, OCU466 and A5. In conclusion, the current study resulted in the prioritization of vital drug targets, which opens future avenues to perform structural as well as biochemical studies on predicted drug targets against *M. avium* subsp. *hominissuis*.

Keywords: *Mycobacterium avium*; tuberculosis; unique metabolic pathways; subtractive genomics; drug target; uncharacterized proteins

1. Introduction

Mycobacterium species that do not cause tuberculosis are referred to as non-tuberculous mycobacteria (NTM) and are ubiquitous in nature. NTM cause pulmonary diseases in which organisms of *Mycobacterium avium* complex (MAC) are widely distributed [1]. The incidence rate of infection caused by *M. avium* is found to be higher than that of the other *Mycobacterium* species. For example, a literature survey showed that the pulmonary infection rate in Japan is sevenfold greater by *M. avium* than any other *Mycobacterium* species [2]. MAC consists of two closely linked species, *M. intracellulare* and *M. avium* [3]. Furthermore, *M. avium* is comprised of four subspecies: *M. avium* subsp. *paratuberculosis* (MAP), *M. avium* subsp. *avium* (MAA), *M. avium* subsp. *silvaticum* (MAS) and

M. avium subsp. *hominissuis* (MAH); and each one is host specific. The first two subspecies cause avian infection, while the third causes diseases in wild livestock and the last one is the most common pathogen in humans and other mammals, including pigs, and therefore has huge economic impact [4].

Opportunistic MAH is responsible for causing disseminated and pulmonary infections that affect immunocompromised patients who are suffering from AIDS, leukemia, lung diseases or chemotherapy [5,6]. The bacterial virulence factor and host-related risk factor contribute to MAC pulmonary diseases. The prevalence of the disease is relatively high in women; however, much of the information about the bacterial virulence factor is still unknown [7]. Environmental risk factors also arise when patients with MAC pulmonary disease are exposed to soil at home or in soil pots [8]. The disease is characterized by adherence to the respiratory mucosa, formation of biofilms [9] and lesions in the linings of epithelial cells of the lungs [7].

MAC pulmonary diseases are controlled by treatment with antibiotics that include macrolide-based multidrug therapy, comprising macrolides (clarithromycin or azithromycin) in combination with rifampin, ethambutol, aminoglycosides (streptomycin or amikacin) and ciprofloxacin [10,11]. However, emerging virulent strains are found to be resistant to these antibiotics [12]. Consequently, these life-threatening microbial pathogens pose an alarming threat for scientists to combat emerging antibiotics resistance. In fact, the emerging strains are capable of becoming more virulent and tolerant to existing drugs [13]. However, the application of genomics has brought about a revolution in the field of drug discovery by providing increased information about the microbial as well as the human genome [14]. This genomic information unveils the mechanism through which pathogens cause the infection. Finding novel and unique drug targets is one of the possible and alternative approaches to overcoming the infections caused by such drug-resistant pathogens. Similarly, finding therapeutic drugs to combat infections of lethal organisms is the most widely applied method albeit with limited success with respect to drug-resistant pathogens [15]. In this scenario, advancements in the fields of computational biology and bioinformatics tools paved the way to propose new and unique drug targets using the subtractive genomics strategy. In the subtractive genomics approach, the genomes of the host and the pathogen are compared, and the non-host pathogen's unique and essential proteins are proposed as drug targets that are vital to the pathogen's survival [16,17]. This strategy recognizes genes that are absent in the host, so called "non-host" genes; however, these genes must be present in the pathogen for its survival, replication and sustainability. Additionally, these non-host genes play crucial roles in unique metabolic pathways and mechanisms. Therefore, when the pathogen's metabolic targets are ideally hit by therapeutic compounds, the therapy must affect the function of the pathogen without altering the host biology [18,19]. The disruption of the essential genes will eventually overcome the pathogen's infection. Recently, several studies applied the same approach for the identification of potential drug targets of *Acinetobacter baumannii* [20], *Helicobacter pylori* [21], *Mycobacterium* species [22], *Pseudomonas aeruginosa* [23] and others [24–28]. Such computational studies help to minimize experimental efforts with high-speed performance for the prioritization of drug targets. For example, by using the information retrieved from such computational studies, a life scientist can express only the prioritized target gene (which is predicted as a potential drug target), resulting in saving the cost of extra experiments and fostering the research.

2. Results and Discussion

With the aim to identify unique and potential druggable targets of *M. avium* subsp. *hominissuis* (MAH), the subtractive genomics method was used, which is the most applicable approach to prioritize potential drug targets [18,29–31].

2.1. Removal of Duplicate Sequences after Proteome Retrieval

Three strains of MAH, i.e., MAH-TH135, OCU466 and A5, were selected from the available non-redundant strains of *M. avium* subsp. *hominissuis* in the UniProt database. Their complete proteomes were downloaded in FASTA format in February 2019. On applying CD-HIT algorithm with 80% identity, 20 sequences were identified as paralogous out of 4614 proteins in MAH-TH135, 54 out of 5165 in MAH-OCU466 and 14 out of 4502 proteins of A5 strain. The CD-HIT clustered the paralogous sequences and, hence, reduced the total number of sequences of each strain. The sequence dataset was comprised of 4596, 5111 and 4488 protein sequences for the MAH-TH135, OCU466 and A5 strains, respectively.

2.2. *Searching of Essential, Non-Homologous and Druggable Proteins*

In this step, protein sequences that were only present in the pathogens were segregated. Thus, by applying a subtractive approach, sequences were excluded that showed similarity to the human host. The remaining orthologous sequences, retrieved from the previous step, were subjected to BLASTp against the complete human proteome, and the resultant file was parsed. The only sequences that were retained were those that showed "no hits found", and a total of 3151, 3619 and 3072 non-homologous sequences were found in the MAH-TH135, OCU466 and A5 strains, respectively.

The Database of Essential Genes (DEG) provides information on essential genes of Gram-positive and Gram-negative bacteria determined from experimental methods (http://www.essentialgene.org/). Homology with the sequences found in the DEG database is the basis of essentiality of non-homologous proteins. To do this, the parsed results of each strain from the last step were subjected to BLASTp against the DEG with a 10^{-5} threshold. The BLASTp results depict 1360, 1451 and 1352 essential protein sequences in MAH-TH135, OCU466 and A5, respectively. These identified sequences were considered viable for the pathogen's life cycle. These sequences include functional, non-functional or uncharacterized proteins, and they were dealt with using different bioinformatics tools for further characterization.

2.3. *Characterization of Essential Non-Homologous Proteins*

2.3.1. Subcellular Localization

The tracing of the location of essential proteins is an important facet to understand the functions of proteins in their suitable cell compartments. It is important to know the localization of a drug target in order to optimize the mode of action of the drug for its specific target. The prediction of sub-cellular localization of the essential non-homologous protein sequences was achieved by a computational tool called PSORTb. The results depict that approximately 48% of proteins resided in the cytoplasm of each strain. A proportion of 23% was distributed in the cytoplasmic membrane. The rest of the proteins were present in different regions, including ~1% of proteins in the extracellular region, >1.5% proteins in the periplasm and very few proteins in the outer membrane of each of the strains. Despite these results, some fractions were considered "unknown" due to the tool's prediction of proteins in multiple sites simultaneously. The distribution of proteins by PSORTb is graphically shown for each strain in Figure 1.

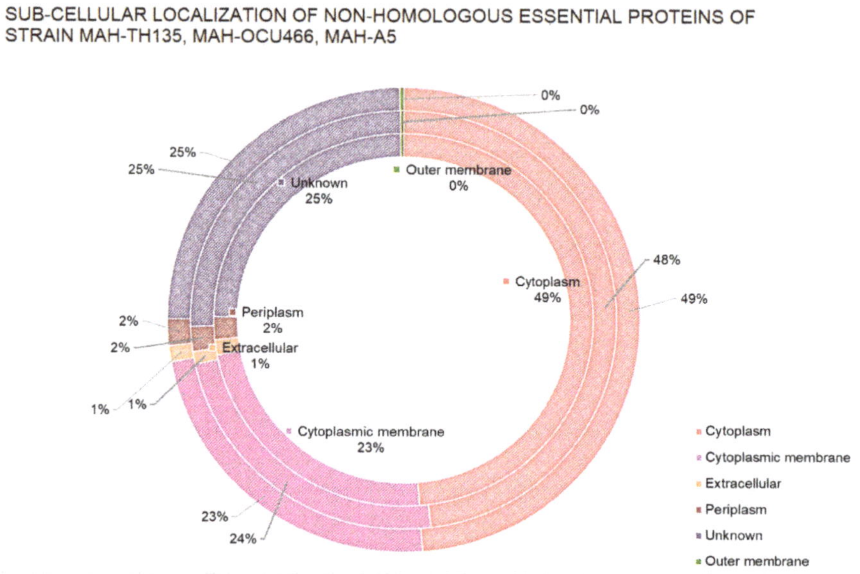

Figure 1. Sub-cellular localization of non-homologous essential proteins. The outermost circle refers to strain MAH-TH135, the middle circle represents strain OCU-466 and the inner circle denotes strain A5.

2.3.2. Functional Family Classification

The functional families of protein sequences were also determined using the Support Vector Machine of Proteins (SVM-Prot) tool. Only the sequences whose functions were not known earlier were submitted to this tool. Hence, only uncharacterized sequences were retrieved from the non-homologous essential proteins' sequences. About 193, 119 and 187 uncharacterized sequences of TH135, OCU466 and A5 strains, respectively, were predicted by the SVM-Prot method. The results of the SVM-Prot tool are depicted in Figure 2. The proteins were broadly classified based on their molecular and biological functions and were further sub-divided into several protein classes, i.e., enzymes, transporters, trans-membranes, zinc or magnesium binding or other elements, DNA condensation, repair, etc. Complete information on classes with their strains is summarized in Supplementary Table S1.

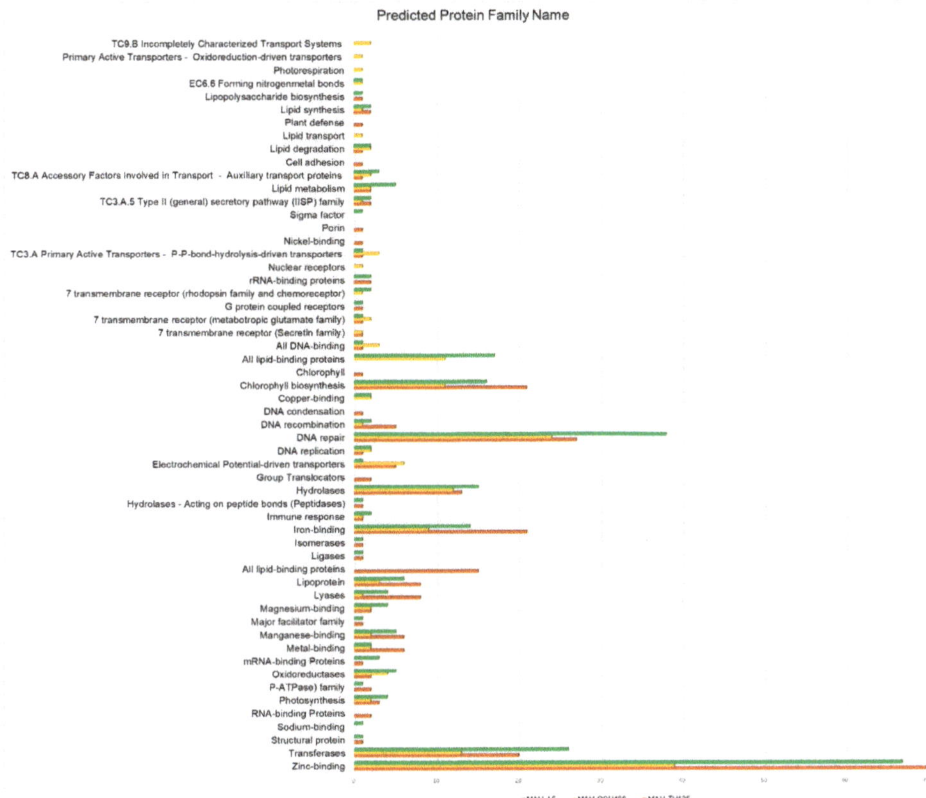

Figure 2. Functional family prediction of *M. avium* subsp. *Hominissuis* (MAH) strains by the SVM-Prot method. The x-axis reports the frequency of each protein family.

2.3.3. Metabolic Pathway Analysis via KEGG

The KEGG database provides a network of metabolic pathways with their complete annotation. It helps to predict which protein sequences are essential in playing a unique role in metabolism. This step predicts the potential drug target based on the pathogen's unique metabolism. Metabolic pathways analysis was carried out for the essential protein sequences using the KEGG database. The DEG's results were subjected to the KEGG database via the KEGG Automated Annotation Server (KAAS). Briefly, out of 675 protein sequences of the MAH-135 strain, 72, 70, 29, 16 and 103 proteins were found to take part in carbohydrate metabolism, energy metabolism, lipid metabolism, nucleotide metabolism and amino acid metabolism, respectively. For OCU-466, 76 were involved in carbohydrate metabolism, while 69, 30 and 15 took part in energy metabolism, lipid metabolism and nucleotide metabolism, respectively, whereas the A5 strain possessed 93 proteins that majorly contributed to amino acid metabolism. The distribution of proteins in different metabolisms is presented in Figure 3a–c. Details are provided in Supplementary Tables S2–S4.

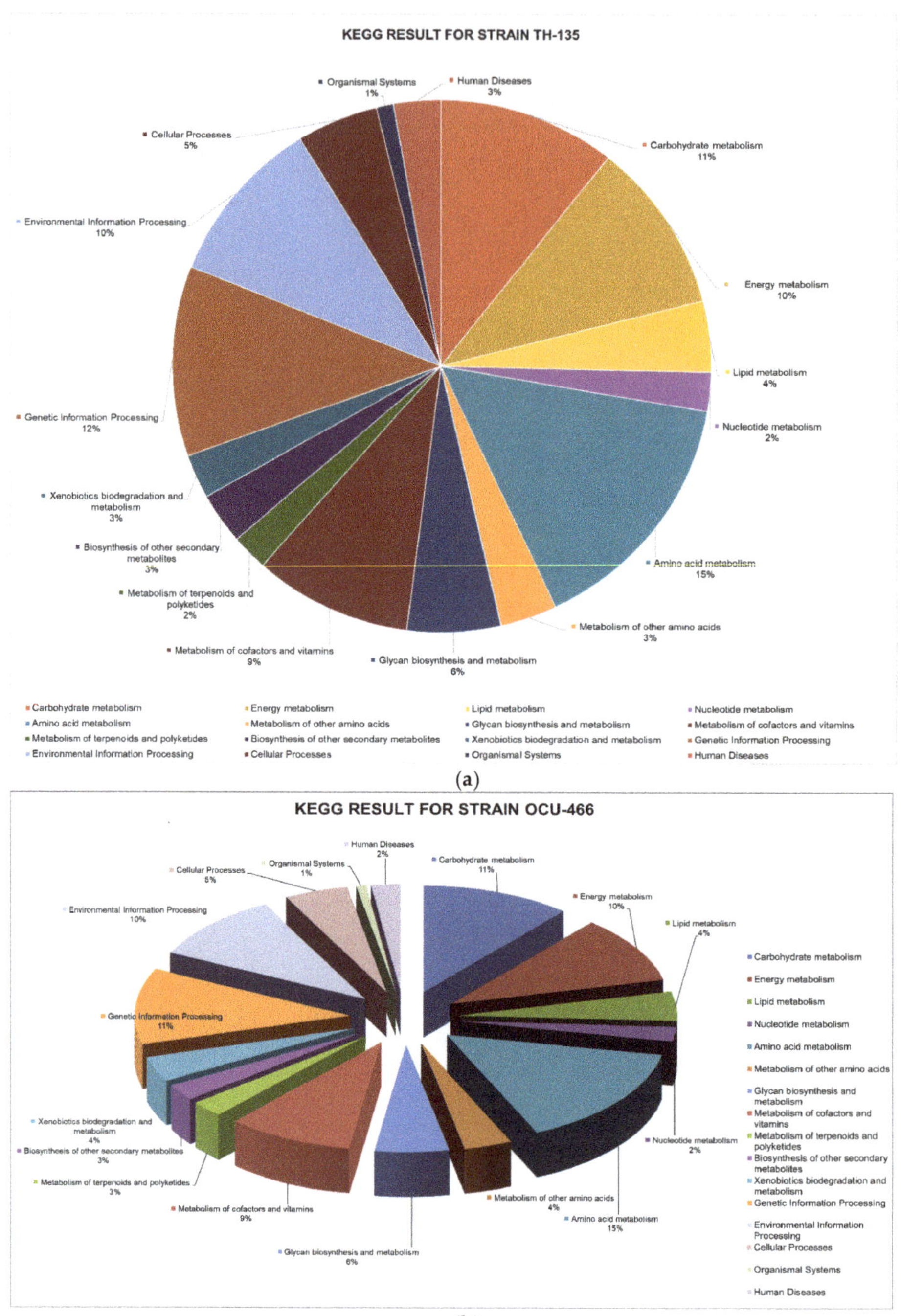

(a)

(b)

Figure 3. *Cont.*

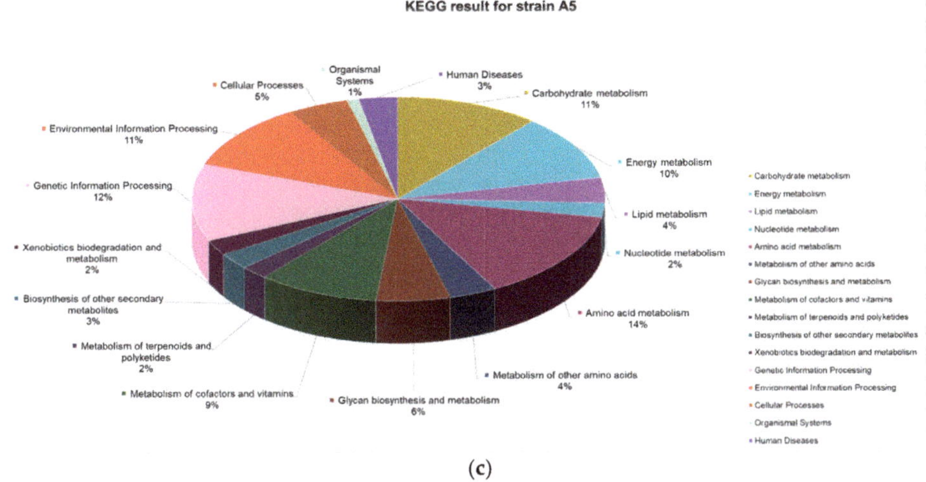

(c)

Figure 3. Percent distribution of non-homologous essential proteins involved in different metabolic pathways of the (**a**) MAH-TH135, (**b**) MAH-OCU466 and (**c**) MAH-A5 strains.

2.4. Discussion of Significant Unique Metabolic Pathways (UMPs) of the Pathogens

Bacterial metabolism refers to the collection of the biochemical reactions required for bacterial survival and growth, which mainly includes respiration (aerobic and anaerobic) and fermentation. Bacteria, as a pathogen to humans, conduct all the same types of basic biochemical reactions a human cell performs. However, bacteria may have several types of energy generating metabolisms that do not exist in human or eukaryotic cells. This diversity of energy generation and metabolism allows bacteria to survive in a variety of habitats and flourish in otherwise not-suitable conditions. On the other hand, these differential metabolic pathways make bacteria susceptible by serving as an ideal target for antibiotics. Metabolic pathways that exist only in pathogens are called unique metabolic pathways (UMP). These UMPs are listed in Supplementary Table S5. We provide brief information on some bacterial UMPs and their significance as an antibiotic target.

2.4.1. Energy Metabolism

Energy is a potential, needed to perform work and maintain life, usually acquired by breaking a chemical bond and stored by making another chemical bond, very often in the form of ATP. Methane metabolism is one of the UMPs by which bacteria can obtain energy by oxidizing one-carbon compounds (e.g., methanol, methane). Methanotrophic bacteria are generally considered environmentally friendly organisms, as they contribute to oxidizing environmental methane, thereby mitigating the effects of global warming [32]. Methane monooxygenases are the main enzymes to catalyze methane oxidation [33]. There are several UMPs in bacteria, which are related to photosynthesis and carbon fixation and can be exploited for the purpose of drug target identification.

2.4.2. Biosynthesis of Secondary Metabolites

Secondary metabolites are molecules not essentially required for the survival of an organism. A large portion of bacterial metabolism deals with the biosynthesis of secondary metabolites. However, these pathways have a minimal role in bacterial growth and viability and are not considered a suitable target for antibiotics. Even though secondary metabolites are not considered to be ideal as drug targets, many of these pathways are manipulated by researchers for valuable purposes such as penicillin and cephalosporin biosynthesis, carbapenem biosynthesis and streptomycin biosynthesis.

2.4.3. Amino Acid Metabolism

Amino acid metabolism in bacteria is diverse in nature and performs a pivotal role in maintaining bacterial growth. Amino acid metabolism has emerged as a potential target for new antibiotics, and a number of new drug targets have been proposed in recent years [34–37]. Some of these drug targets have shown promising results. Lysine biosynthesis, an essential pathway in bacteria for survival and growth, is reported to be a potential target for antibiotics [38,39]. Similarly, *D*-alanine metabolism is a significant target; an antibiotic *D*-cycloserine targeting *D*-alanine metabolism is already in clinical use against *Mycobacterium tuberculosis* [40,41]. The heterogeneity of amino acid metabolism implies an enormous scope for discovering new antibiotic targets using modern computational tools.

Other types of metabolic activities in bacteria, such as terpenoids and polyketides, glycan biosynthesis and drug resistance, also perform supportive functions for bacterial growth and survival; however, these metabolic routes are not prioritized targets for anti-bacterial drugs. Rather, these metabolic routes are often manipulated for advantageous purposes [42].

2.5. Shortlisting of Proteins Sequences as Druggable

The potential drug targets were shortlisted based on obtained information from earlier successful literature reports. The druggability of non-host uncharacterized protein sequences was determined by performing BLASTp against the druggable protein sequences present in the DrugBank Database. For this purpose, the earlier shortlisted, non-host, uncharacterized proteins, which are essential in metabolic pathways, were analyzed for druggability by comparing their sequences with the DrugBank Database. In this search, only one protein was prioritized in TAH-135, whereas four and seven potential drug targets emerged with the OCU-466 and A5 strains, respectively (Table 1). All these potential drug targets were similar to the FDA-approved drug target sequences in the DrugBank Database, including the DNA polymerase III subunit ε of the TH-135 strain, Inter-α-trypsin inhibitor heavy chain H4, exopolyphosphatase, DNA polymerase III subunit ε, mannoside ABC transport system and sugar-binding protein of the OCU-466 strain. In addition to all the proteins from the OCU-466 strain, diacylglycerol acyltransferase/mycolyltransferase, Ag85C and nickel-binding periplasmic protein were found for the A5 strain.

Table 1. Protein drug targets of *M. avium* subsp. *hominissuis*.

		UNIPROT STRAIN ID MAH-TH135		
S. No.	**Protein ID**	**DrugBank target name**	**DrugBank ID**	**Localization Site**
1.	T2GUW6	DNA polymerase III subunit epsilon (DB01643)	P03007	Cytoplasmic
		UNIPROT STRAIN ID MAH-OCU466		
S. No.	**Protein ID**	**DrugBank target name**	**DrugBank ID**	**Localization Site**
1.	A0A2A3L1J8	Inter-alpha-trypsin inhibitor heavy chain H4 (DB01593; DB14487; DB14533) Inter-alpha-trypsin inhibitor heavy chain H4 (DB01593; DB14487; DB14533)	Q14624 Q06033	Cytoplasmic
2.	A0A2A3L805	O67040 Exopolyphosphatase (DB03382)	O67040	Cytoplasmic
3.	A0A2A3L3Y2	DNA polymerase III subunit epsilon (DB01643)	P03007	Cytoplasmic
4.	A0A2A3LDY9	Mannoside ABC transport system, sugar-binding protein (DB01942)	Q9X0V0	Unknown

Table 1. *Cont.*

UNIPROT STRAIN ID MAH-A5				
S. No.	**Protein ID**	**DrugBank target name**	**DrugBank ID**	**Localization Site**
1.	A0A0E2W125	Exopolyphosphatase (DB03382)	O67040	Cytoplasmic
2.	A0A0E2W9K2	Inter-alpha-trypsin inhibitor heavy chain H4 (DB01593; DB14487; DB14533)	Q14624	Cytoplasmic
		Inter-alpha-trypsin inhibitor heavy chain H4 (DB01593; DB14487; DB14533)	Q06033	
3.	A0A0E2W6U1	Diacylglycerol acyltransferase/mycolyltransferase Ag85C (DB02811; DB08558)	P9WQN9	Unknown (This protein may have multiple localization sites.)
4.	A0A0E2W8I5	Diacylglycerol acyltransferase/mycolyltransferase Ag85C (DB02811; DB08558)	P9WQN9	Extracellular
5.	A0A0E2W8U0	DNA polymerase III subunit epsilon (DB01643)	P03007	Cytoplasmic
6.	A0A0E2WAR7	Mannoside ABC transport system, sugar-binding protein (DB01942)	Q9X0V0	Unknown
		Nickel-binding periplasmic protein (DB03374)	P33590	
7.	A0A0E2WQA2	Mannoside ABC transport system, sugar-binding protein (DB01942)	Q9X0V0	Periplasmic
		Nickel-binding periplasmic protein (DB03374)	P33590	
		Periplasmic oligopeptide-binding protein (DB07365)	P06202	
		ABC transporter, periplasmic substrate-binding protein (DB02078)	Q5LRQ9	

It is noteworthy that all the proposed drug targets could be analyzed for 3D structural information to prioritize novel drug targets against pathogens. Therefore, BLASTp was performed for the target proteins against the Protein Data Bank (PDB) database, which revealed that 12 protein sequences had no 3D structure available yet in the PDB. Therefore, this study offers those 12 proteins' sequences to not only consider as a potential druggable genome, but also for future studies of 3D structure determination either by homology modeling (template-based) or by ab initio (template-free) methods [43].

3. Materials and Methods

An overview of the subtractive genomics approach is illustrated in Figure 4.

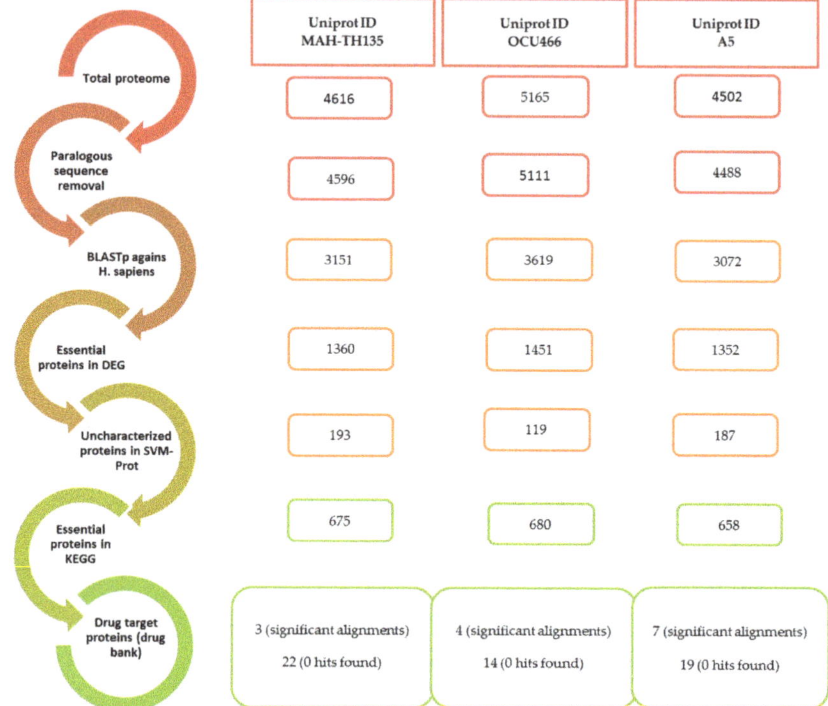

Figure 4. Workflow of the subtractive genomics approach.

3.1. Extraction of the Host–Pathogen Proteome

The whole proteome of the host, i.e., *Homo sapiens*, and pathogen, i.e., *Mycobacterium avium* subsp. *Hominissuis*, were downloaded from the UniProt KB database [44] to retrieve protein sequences. The drug target identification approach was carried out on the pathogenic MAH-TH135, MAH-OCU466 and A5 strains.

3.2. Grouping of Common Proteins in All Strains

The CD-HIT tool [45] clusters the protein or nucleotide sequences and reduces redundancy and manual efforts in sequence analysis. This tool was used as a standalone command line tool to remove paralogous or duplicated sequences of all strains with a threshold value of 80%. The remaining set of proteins was grouped as orthologous sequences.

3.3. Identification of Non-Homologous Proteins

Standalone BLAST version 2.8.1 was downloaded from the NCBI FTP server [46]. The orthologous sequences were subjected to BLASTp against the *H. sapiens* database with an expectation value (e-value) of 10^{-3} [47]. The output was obtained with keywords of "no hits found" for unique proteins and "significant alignments" for the sequences having similarity with the human (host) proteome. The results were analyzed, and only protein sequences "with no homology with the human host" were retained, while the rest were removed. Those proteins were further labelled as non-homologous proteins, and finally, they were extracted using our in-house scripts.

3.4. Finding of Essential Genes

The genes required to sustain the life cycle of bacteria are called essential genes. The Database of Essential Genes (DEG) contains lists of genes with their corresponding sequences, which are essential for the survival of bacterial life. [48]. Therefore, the DEG was used to find the sequences that are essential to the bacterial pathogen studied here (i.e., *M. avium* subsp. *hominissuis*). The non-homologous proteins were aligned with the DEG database using BLASTp, and the expectation value was set to 10^{-5}. As a result, the non-homologous essential genes, which may have hypothetical or uncharacterized proteins, were obtained.

3.5. Information about Metabolic Pathways

The metabolic pathways of the identified non-homologous essential proteins were searched in the Kyoto Encyclopedia of Genes and Genomes (KEGG) [49] through the KAAS server. KAAS [50] uses BLASTp for the comparison of query proteins against the KEGG database and annotates functions. KAAS provides the KEGG Orthology (KO) identifiers and information on the metabolic pathways of the proteins.

3.6. Annotation of the Curated Proteins

Annotation of proteins includes information about the location of proteins in various regions of the cell and the family to which it belongs. PSORTb version 3.0 [51] is well known to predict the subcellular localization (SCL) of proteins. The SCL includes different compartments, such as cytoplasmic membrane, cytoplasm, cell wall and extracellular and unknown regions of the cell where the proteins reside. All the non-homologous essential, as well as hypothetical, proteins were subjected to the protein databases with known functions using SCL BLAST by the web-based server. SVM-Prot [52] is an online tool for the classification of protein functional families. It applies the machine-learning method and predicts a diverse set of molecular and biological functions covering all major classes of enzymes, channels, transporters, receptors, DNA/RNA binding proteins, etc. and covering 192 functional families of proteins. Those proteins whose functions are still unknown were labeled as non-homologous, hypothetical/uncharacterized proteins and passed through the server of SVM-Prot to classify them into functional families.

3.7. Druggability of the Shortlisted Sequences

In order to detemine the novel drug targets, standalone BLASTp was run between hypothetical non-homologous essential proteins, and drug target sequences were taken from the DrugBank Database [53] with an e-value cutoff 10^{-3}. The DrugBank Database provides detailed information on drugs and drug targets. A large database shows up to 8261 drugs, including FDA-approved drugs; experimental and nutraceutical drugs are available in the DrugBank Database.

4. Conclusions

Different bioinformatics tools were applied in this study to identify vital drug targets of *Mycobacterium avium* subsp. *hominissuis*. Protein sequences of *M. avium* subsp. *hominissuis* were parsed using multiple steps of the subtractive genomics approach, and a few of them were shortlisted as possible drug targets because they fulfilled the druggability criteria. The shortlisted sequences were non-homologous to the human host; thus, these can be proposed as ideal drug targets. All the identified drug targets of different strains of MAH have never been characterized before as drug targets, and we proposed them here as potential drug targets against which new drug compounds can be designed. Therefore, the study is significant to the scientific community, as it provides a prioritized list of possible drug targets sorted by the computational subtractive genomics method, and it has the potential to lead to the discovery of new and novel drug targets against *M. avium* subsp. *hominissuis*.

Supplementary Materials: The following are available online at http://www.mdpi.com/2076-0817/9/5/368/s1, Table S1; Table S2; Table S3; Table S4; Table S5.

Author Contributions: Conceptualization, R.U. and A.A.-H.; methodology, BS.; software, M.R.; formal analysis, B.S.; investigation, R.U and A.K.; resources, R.U.; data curation, S.A.H.; writing—original draft preparation, R.U.; writing—review and editing, A.K.; visualization, A.K. and S.A.H.; supervision, A.A.-H.; funding acquisition, A.A.-H. All authors have read and agreed to the published version of the manuscript.

Funding: This research was funded by Pakistan Science Foundation Grant# PSF-TUBITAK/S-HEJ (04) and the APC was funded by University of Nizwa.

Acknowledgments: The authors would gratefully like to acknowledge the financial support provided by the Pakistan Science Foundation Grant# PSF-TUBITAK/S-HEJ (04) for this project.

Conflicts of Interest: The authors declare no conflicts of interest.

References

1. Daley, C. Mycobacterium avium Complex Disease. *Microbiol. Spectr.* **2017**, *5*, 663–701. [CrossRef]
2. Iwamoto, T.; Nakajima, C.; Nishiuchi, Y.; Kato, T.; Yoshida, S.; Nakanishi, N.; Tamaru, A.; Tamura, Y.; Suzuki, Y.; Nasu, M. Genetic diversity of *Mycobacterium avium* subsp. *hominissuis* strains isolated from humans, pigs, and human living environment. *Infect. Genet. Evol.* **2012**, *12*, 846–852. [CrossRef]
3. Uchiya, K.-I.; Takahashi, H.; Yagi, T.; Moriyama, M.; Inagaki, T.; Ichikawa, K.; Nakagawa, T.; Nikai, T.; Ogawa, K. Comparative genome analysis of *Mycobacterium avium* revealed genetic diversity in strains that cause pulmonary and disseminated disease. *PLoS ONE* **2013**, *8*, e71831. [CrossRef] [PubMed]
4. Mijs, W.; De Haas, P.; Rossau, R.; Van Der Laan, T.; Rigouts, L.; Portaels, F.; Van Soolingen, D. Molecular evidence to support a proposal to reserve the designation *Mycobacterium avium* subsp. avium for bird-type isolates and '*M. avium* subsp. *hominissuis*' for the human/porcine type of *M. avium*. *Int. J. Syst. Evol. Micr.* **2002**, *52*, 1505–1518.
5. Porvaznik, I.; Solovič, I.; Mokrý, J. Non-tuberculous mycobacteria: Classification, diagnostics, and therapy. In *Respiratory Treatment and Prevention*; Springer: Berlin/Heidelberg, Germany, 2016; pp. 19–25.
6. Bruffaerts, N.; Vluggen, C.; Roupie, V.; Duytschaever, L.; Van den Poel, C.; Denoël, J.; Wattiez, R.; Letesson, J.-J.; Fretin, D.; Rigouts, L. Virulence and immunogenicity of genetically defined human and porcine isolates of *M. avium* subsp. *hominissuis* in an experimental mouse infection. *PLoS ONE* **2017**, *12*, e0171895. [CrossRef] [PubMed]
7. Uchiya, K.-I.; Takahashi, H.; Nakagawa, T.; Yagi, T.; Moriyama, M.; Inagaki, T.; Ichikawa, K.; Nikai, T.; Ogawa, K. Characterization of a novel plasmid, pMAH135, from *Mycobacterium avium* subsp. *hominissuis*. *PLoS ONE* **2015**, *10*, e0117797. [CrossRef]
8. Maekawa, K.; Ito, Y.; Hirai, T.; Kubo, T.; Imai, S.; Tatsumi, S.; Fujita, K.; Takakura, S.; Niimi, A.; Iinuma, Y. Environmental risk factors for pulmonary *Mycobacterium avium*-intracellulare complex disease. *Chest* **2011**, *140*, 723–729. [CrossRef]
9. Weiss, C.; Glassroth, J. Pulmonary disease caused by nontuberculous mycobacteria. *Expert. Rev. Respir. Med.* **2012**, *6*, 597–613. [CrossRef]
10. Uchiya, K.-I.; Asahi, S.; Futamura, K.; Hamaura, H.; Nakagawa, T.; Nikai, T.; Ogawa, K. Antibiotic susceptibility and genotyping of *Mycobacterium avium* strains that cause pulmonary and disseminated infection. *Antimicrob. Agents Chemother.* **2018**, *62*, e02035-17. [CrossRef]
11. Blanchard, J.D.; Elias, V.; Cipolla, D.; Gonda, I.; Bermudez, L.E. Effective Treatment of *Mycobacterium avium* subsp. hominissuis and *Mycobacterium abscessus* Species Infections in Macrophages, Biofilm, and Mice by Using Liposomal Ciprofloxacin. *Antimicrob. Agents Chemother.* **2018**, *62*, e00440-18. [CrossRef]
12. Griffith, D.E.; Brown-Elliott, B.A.; Langsjoen, B.; Zhang, Y.; Pan, X.; Girard, W.; Nelson, K.; Caccitolo, J.; Alvarez, J.; Shepherd, S. Clinical and molecular analysis of macrolide resistance in *Mycobacterium avium* complex lung disease. *Am. J. Respir. Crit. Care Med.* **2006**, *174*, 928–934. [CrossRef] [PubMed]
13. Nicolle, L. Community-acquired MRSA: A practitioner's guide. *CMAJ* **2006**, *175*, 145. [CrossRef] [PubMed]
14. Rathi, B.; Sarangi, A.N.; Trivedi, N. Genome subtraction for novel target definition in *Salmonella typhi*. *Bioinformation* **2009**, *4*, 143–150. [CrossRef] [PubMed]

15. Butt, A.M.; Tahir, S.; Nasrullah, I.; Idrees, M.; Lu, J.; Tong, Y. Mycoplasma genitalium: A comparative genomics study of metabolic pathways for the identification of drug and vaccine targets. *Infect. Genet. Evol.* **2012**, *12*, 53–62. [CrossRef]
16. Barh, D.; Tiwari, S.; Jain, N.; Ali, A.; Santos, A.R.; Misra, A.N.; Azevedo, V.; Kumar, A. *In silico* subtractive genomics for target identification in human bacterial pathogens. *Drug Develop. Res.* **2011**, *72*, 162–177. [CrossRef]
17. Bottacini, F.; Motherway, M.O.C.; Kuczynski, J.; O'Connell, K.J.; Serafini, F.; Duranti, S.; Milani, C.; Turroni, F.; Lugli, G.A.; Zomer, A. Comparative genomics of the Bifidobacterium breve taxon. *BMC Genom.* **2014**, *15*, 170. [CrossRef]
18. Uddin, R.; Saeed, K. Identification and characterization of potential drug targets by subtractive genome analyses of methicillin resistant *Staphylococcus aureus*. *Comput. Biol. Chem.* **2014**, *48*, 55–63. [CrossRef]
19. Galperin, M.Y.; Koonin, E.V. Searching for drug targets in microbial genomes. *Curr. Opin. Biotechnol.* **1999**, *10*, 571–578. [CrossRef]
20. Uddin, R.; Masood, F.; Azam, S.S.; Wadood, A. Identification of putative non-host essential genes and novel drug targets against *Acinetobacter baumannii* by *in silico* comparative genome analysis. *Microb. Pathog.* **2019**, *128*, 28–35. [CrossRef]
21. Dutta, A.; Singh, S.K.; Ghosh, P.; Mukherjee, R.; Mitter, S.; Bandyopadhyay, D. *In silico* identification of potential therapeutic targets in the human pathogen *Helicobacter pylori*. *Silico Biol.* **2006**, *6*, 43–47.
22. Marri, P.R.; Bannantine, J.P.; Golding, G.B. Comparative genomics of metabolic pathways in Mycobacterium species: Gene duplication, gene decay and lateral gene transfer. *FEMS Microbiol. Rev.* **2006**, *30*, 906–925. [CrossRef] [PubMed]
23. Uddin, R.; Jamil, F. Prioritization of potential drug targets against *P. aeruginosa* by core proteomic analysis using computational subtractive genomics and protein-Protein interaction network. *Comput. Biol. Chem.* **2018**, *74*, 115–122. [CrossRef] [PubMed]
24. Ahmad, S.; Navid, A.; Akhtar, A.S.; Azam, S.S.; Wadood, A.; Pérez-Sánchez, H. Subtractive Genomics, Molecular Docking and Molecular Dynamics Simulation Revealed LpxC as a Potential Drug Target Against Multi-Drug Resistant *Klebsiella pneumoniae*. *Interdiscipl. Sci. Comput. Life Sci.* **2019**, *11*, 508–526. [CrossRef] [PubMed]
25. Asalone, K.C.; Nelson, M.M.; Bracht, J.R. Novel Sequence Discovery by Subtractive Genomics. *J. Vis. Exp.* **2019**, *143*, e58877. [CrossRef] [PubMed]
26. Nayak, S.; Pradhan, D.; Singh, H.; Reddy, M.S. Computational screening of potential drug targets for pathogens causing bacterial pneumonia. *Microb. Pathog.* **2019**, *130*, 271–282. [CrossRef]
27. Prabha, R.; Singh, D.P.; Ahmad, K.; Kumar, S.P.J.; Kumar, P. Subtractive genomics approach for identification of putative antimicrobial targets in *Xanthomonas oryzae* pv. oryzae KACC10331. *Arch. Phytopath. Plant Protect.* **2019**, *52*, 863–872. [CrossRef]
28. Auster, L.; Sutton, M.; Gwin, M.C.; Nitkin, C.; Bonfield, T.L. Optimization of *In Vitro Mycobacterium avium* and Mycobacterium intracellulare Growth Assays for Therapeutic Development. *Microorganisms* **2019**, *7*, 42. [CrossRef]
29. Shoukat, K.; Rasheed, N.; Sajid, M. Subtractive genome analysis for *In silico* identification and characterization of novel drug targets IN C. trachomatis STRAIN D/UW-3/Cx. *Int. J. Curr. Res.* **2012**, *4*, 017–021.
30. Koteswara, R.G.; Nagamalleswara, R.K.; Phani, R.; Krishna, B.; Aravind, S. *In silico* identification of potential therapeutic targets *inclostridium botulinum* by the approach subtractive genomics. *Int. J. Bioinform. Res.* **2010**, *2*, 12–16.
31. Sharma, V.; Gupta, P.; Dixit, A. *In silico* identification of putative drug targets from different metabolic pathways of *Aeromonas hydrophila*. *Silico Biol.* **2008**, *8*, 331–338.
32. Hanson, R.S.; Hanson, T.E. Methanotrophic bacteria. *Microbiol. Rev.* **1996**, *60*, 439–471. [CrossRef] [PubMed]
33. Dalton, H. Structure and Mechanism of Action of the Enzyme(s) Involved in Methane Oxidation. In *Applications of Enzyme Biotechnology*; Kelly, J.W., Baldwin, T.O., Eds.; Springer: Boston, MA, USA, 1991; pp. 55–68.
34. Beste, D.J.; Noh, K.; Niedenfuhr, S.; Mendum, T.A.; Hawkins, N.D.; Ward, J.L.; Beale, M.H.; Wiechert, W.; McFadden, J. 13C-flux spectral analysis of host-pathogen metabolism reveals a mixed diet for intracellular *Mycobacterium tuberculosis*. *Chem. Biol.* **2013**, *20*, 1012–1021. [CrossRef] [PubMed]

35. Gouzy, A.; Larrouy-Maumus, G.; Bottai, D.; Levillain, F.; Dumas, A.; Wallach, J.B.; Caire-Brandli, I.; De Chastellier, C.; Wu, T.D.; Poincloux, R.; et al. *Mycobacterium tuberculosis* exploits asparagine to assimilate nitrogen and resist acid stress during infection. *PLoS Pathog.* **2014**, *10*, e1003928. [CrossRef] [PubMed]
36. Gouzy, A.; Larrouy-Maumus, G.; Wu, T.D.; Peixoto, A.; Levillain, F.; Lugo-Villarino, G.; Guerquin-Kern, J.L.; De Carvalho, L.P.; Poquet, Y.; Neyrolles, O. *Mycobacterium tuberculosis* nitrogen assimilation and host colonization require aspartate. *Nat. Chem. Biol.* **2013**, *9*, 674–676. [CrossRef]
37. Tullius, M.V.; Harth, G.; Horwitz, M.A. Glutamine synthetase GlnA1 is essential for growth of *Mycobacterium tuberculosis* in human THP-1 macrophages and guinea pigs. *Infect. Immun.* **2003**, *71*, 3927–3936. [CrossRef]
38. Gillner, D.M.; Becker, D.P.; Holz, R.C. Lysine biosynthesis in bacteria: A metallodesuccinylase as a potential antimicrobial target. *J. Biol. Inorg. Chem* **2013**, *18*, 155–163. [CrossRef]
39. Mandal, R.S.; Das, S. In silico approach towards identification of potential inhibitors of *Helicobacter pylori* DapE. *J. Biomol. Struct. Dyn.* **2015**, *33*, 1460–1473. [CrossRef]
40. Halouska, S.; Fenton, R.J.; Zinniel, D.K.; Marshall, D.D.; Barletta, R.G.; Powers, R. Metabolomics analysis identifies D-Alanine-D-Alanine ligase as the primary lethal target of D-Cycloserine in mycobacteria. *J. Proteome Res.* **2014**, *13*, 1065–1076. [CrossRef]
41. Qiu, W.; Zheng, X.; Wei, Y.; Zhou, X.; Zhang, K.; Wang, S.; Cheng, L.; Li, Y.; Ren, B.; Xu, X.; et al. D-Alanine metabolism is essential for growth and biofilm formation of *Streptococcus mutans*. *Mol. Oral Microbiol.* **2016**, *31*, 435–444. [CrossRef]
42. Silver, L.L. Appropriate Targets for Antibacterial Drugs. *Cold Spring Harb. Perspect. Med.* **2016**, *6*, a030239. [CrossRef]
43. Caffrey, C.R.; Rohwer, A.; Oellien, F.; Marhöfer, R.J.; Braschi, S.; Oliveira, G.; McKerrow, J.H.; Selzer, P.M. A comparative chemogenomics strategy to predict potential drug targets in the metazoan pathogen, *Schistosoma mansoni*. *PLoS ONE* **2009**, *4*, e4413. [CrossRef]
44. Consortium, U. UniProt: A hub for protein information. *Nucleic Acids Res.* **2015**, *43*, D204–D212. [CrossRef] [PubMed]
45. Fu, L.; Niu, B.; Zhu, Z.; Wu, S.; Li, W. CD-HIT: Accelerated for clustering the next-generation sequencing data. *Bioinformatics* **2012**, *28*, 3150–3152. [CrossRef] [PubMed]
46. Tao, T. *Standalone BLAST Setup for Unix*; National Center for Biotechnology Information: Bethesda, MD, USA, 2008.
47. Kerfeld, C.A.; Scott, K.M. Using BLAST to teach "E-value-tionary" concepts. *PLoS Biol.* **2011**, *9*, e1001014. [CrossRef] [PubMed]
48. Gao, F.; Luo, H.; Zhang, C.-T.; Zhang, R. Gene essentiality analysis based on DEG 10, an updated database of essential genes. In *Gene Essentiality*; Springer: Berlin/Heidelberg, Germany, 2015; pp. 219–233.
49. Kanehisa, M.; Furumichi, M.; Tanabe, M.; Sato, Y.; Morishima, K. KEGG: New perspectives on genomes, pathways, diseases and drugs. *Nucleic Acids Res.* **2016**, *45*, D353–D361. [CrossRef] [PubMed]
50. Moriya, Y.; Itoh, M.; Okuda, S.; Yoshizawa, A.C.; Kanehisa, M. KAAS: An automatic genome annotation and pathway reconstruction server. *Nucleic Acids Res.* **2007**, *35*, W182–W185. [CrossRef] [PubMed]
51. Yu, N.Y.; Wagner, J.R.; Laird, M.R.; Melli, G.; Rey, S.; Lo, R.; Dao, P.; Sahinalp, S.C.; Ester, M.; Foster, L.J. PSORTb 3.0: Improved protein subcellular localization prediction with refined localization subcategories and predictive capabilities for all prokaryotes. *Bioinformatics* **2010**, *26*, 1608–1615. [CrossRef] [PubMed]
52. Li, Y.H.; Xu, J.Y.; Tao, L.; Li, X.F.; Li, S.; Zeng, X.; Chen, S.Y.; Zhang, P.; Qin, C.; Zhang, C. SVM-Prot 2016: A web-server for machine learning prediction of protein functional families from sequence irrespective of similarity. *PLoS ONE* **2016**, *11*, e0155290. [CrossRef] [PubMed]
53. Wishart, D.S.; Feunang, Y.D.; Guo, A.C.; Lo, E.J.; Marcu, A.; Grant, J.R.; Sajed, T.; Johnson, D.; Li, C.; Sayeeda, Z. DrugBank 5.0: A major update to the DrugBank database for 2018. *Nucleic Acids Res.* **2017**, *46*, D1074–D1082. [CrossRef]

Article

Interleukin-18, Functional IL-18 Receptor and IL-18 Binding Protein Expression in Active and Latent Tuberculosis

Sebastian Wawrocki [1], Grzegorz Kielnierowski [2], Wieslawa Rudnicka [1], Michal Seweryn [3] and Magdalena Druszczynska [1,*]

[1] Department of Immunology and Infectious Biology, Institute of Microbiology, Biotechnology and Immunology, Faculty of Biology and Environmental Protection, University of Lodz, Banacha 12/16, 90-237 Lodz, Poland; sebastian.wawrocki@biol.uni.lodz.pl (S.W.); wieslawa.rudnicka@biol.uni.lodz.pl (W.R.)

[2] Regional Specialized Hospital of Tuberculosis, Lung Diseases, and Rehabilitation, Szpitalna 5, 95-080 Tuszyn, Poland; kielnier@o2.pl

[3] Center for Medical Genomics OMICRON, Jagiellonian University, Medical College, Swietej Anny 12, 31-008 Cracow, Poland; michal.seweryn@wmii.uni.lodz.pl

* Correspondence: magdalena.druszczynska@biol.uni.lodz.pl

Received: 30 April 2020; Accepted: 4 June 2020; Published: 8 June 2020

Abstract: A thorough understanding of the processes modulating the innate and acquired immune response to *Mycobacterium tuberculosis* (*M.tb*) infection in the context of gene expression is still a scientific and diagnostic problem. The study was aimed to assess IL-18, IL-18 binding protein (IL-18BP), IL-18R, IFN-γ, and IL-37 mRNA expression in patients with active tuberculosis (ATB) and healthy volunteers with latent *M.tb*-infection (LTB) or *M.tb*-uninfected healthy controls (Control). The relative mRNA expression was assessed in the buffy coat blood fraction using the qPCR method. In total, 97 BCG-vaccinated Polish adults were enrolled in the study. The relative expression of IL-18 and IL-18BP mRNA was significantly elevated in the ATB and LTB groups. In ATB, but not LTB individuals, the overexpression of IL-18 and IL-18BP, as well as a significant increase in IFN-γ mRNA expression, might be considered as a manifestation of active tuberculosis disease. No statistically significant differences were observed in the IL-37 mRNA expression among the studied groups. Particularly noteworthy is the outstanding reduction in the relative expression of IL-18R mRNA in the LTB group as compared to the ATB and Control group. Reduced expression of IL-18R in LTB group may, at least partially, prevent the development of a pathological inflammatory reaction and promote the maintenance of homeostatic conditions between host immunity and *M.tb*.

Keywords: tuberculosis; IL-18; IL-18BP; IL-18R; gene expression

1. Introduction

Tuberculosis (TB) is the leading global cause of death from an infectious agent, *Mycobacterium tuberculosis* (*M.tb*). TB affects about 10-million people in the world and is a cause of two-million deaths annually, according to the estimates of the World Health Organization [1]. One-third of the world population carries an asymptomatic *M.tb* infection. These individuals have developed an efficient immune response that allows them to block the metabolic activity of the pathogen, but it does not provide its eradication. People with a latent TB infection (LTB) represent a reservoir of potential progress to disease, because about 5–10% of them will develop active TB disease, if not treated.

The antigen-specific, as well as non-specific, response of the immune cells to *M.tb* infection is modulated by mRNA expression, which results in the production of cytokines and other proteins activating numerous cell populations. Among these, interleukin (IL)-18plays an important role—it induces NK cell cytotoxic activity and promotes the development of Th1 cell response. This mechanism is associated with the production of interferon (IFN)-γ, which is a key element in anti-mycobacterial protection. IL-18 was first described in 1989 as an "IFN-γ inducing factor" [2–4]. Similarly to IL-1β, IL-18 is constantly synthesized as an inactive precursor, and the cysteine protease (caspase-1) is involved in its maturation. In a mouse model, increased susceptibility to infection with *M.tb* was found in animals that were not able to produce IL-18 [5]. Moreover, IL-18 increases the expression of adhesion molecules, the synthesis of enzyme nitric oxide synthase, and the production of chemokines. In addition to inducing a T helper (h) type 1 (Th1) cellular response, IL-18, together with IL-2, leads to a Th2 type cell response and the production of IL-4 and IL-13 [6]. In humans, the gene encoding IL-18 is located on chromosome 11 at position 11q22.2–q22.3 and it consists of six exons. There are several common polymorphisms in the promoter region of IL-18 than affect the transcription factor binding sites and, in turn, might be expression quantitative trait loci (eQTLs) for IL-18. Therefore, these genetic variants may predispose to TB by affecting the expression of the cytokine itself, followed by the development of the IFN-γ-mediated Th1 response [7].

The IL-18 levels are also regulated by soluble IL-18 binding protein (IL-18BP), which is a natural inhibitor of IL-18. Under physiological conditions, the concentration of plasma IL-18BP is ~20 times higher than that of IL-18, which prevents IL-18 from binding to its cellular receptor. The gene coding for IL-18BP is located on chromosome 11 at position 11q13.4. The mRNA promoter region contains two response elements (RE), regulatory sequences sensitive to IFN-γ attachment, which results in increased gene expression and protein production [8,9].

Cell activation by IL-18 occurs via the IL-18R receptor, which belongs to the IL-1 family, the members of which show structural and functional similarity. IL-18R is expressed on many cells, such as macrophages, NK cells, neutrophils, epithelial cells, and smooth muscle. The IL-18R receptor is a heterodimer that is composed of two polypeptide chains: IL-18Rα and IL-18Rβ. The IL-18Rα chain is responsible for ligand binding. However, it binds to IL-18 with low affinity. On the other hand, the IL-18Rβ chain functions as a co-receptor, which enhances the strength with which the receptor binds IL-18 and transmits the signal to the inside of the cell [10–12]. The genes encoding the receptor are located on chromosome 2 at position 2q12.1 There are known genetic variants in the regulatory regions of IL-18R associated genes, which may affect its and functionality [13].

A new member of the IL-1 family, IL-37, has gained increasing attention in recent years. IL-37, to which IL-18BP also has a high affinity, was found to be an important regulator of inflammation [14,15]. IL-37 is also able to bind to IL-18Rα, but with much lower affinity than IL-18 [16]. IL-18BP and IL-37 act to reduce the production of inflammatory cytokines; however, the anti-inflammatory properties of IL-37 depend on the concentration of IL-18 binding protein. The human IL-37 gene cluster is located on chromosome 2 at position 2q14.1 [17].

IFN-γ is a cytokine that is known to be crucial in regulating the immune response to *M.tb* infection. This cytokine is mainly produced by activated CD4 (+) Th1 T lymphocytes, and it also has a key role in inducing nitric oxide (NO)-dependent apoptosis of mycobacteria infected macrophages. IFN-γ was shown to increase the expression of major histocompatibility complex (MHC) class I and II surface molecules and promote differentiation towards Th1 response. Deleterious mutations in the gene encoding the IFN-γ receptor predispose to the acute course of mycobacterial infection [3,18]. The human immune interferon gene is located on chromosome 12at position 12q15 [19]. The expression levels of these cytokines, as well as their mutual relations in sera, as well as in cultures stimulated, were found to be informative of *M.tb* infection status [20,21]. This work aims to assess the level of expression of IL-18, IL-18BP, IL-18R, as well as IFN-γand IL-37 genes in patients with active pulmonary tuberculosis (ATB), healthy individuals with latent *M.tb* infection (LTB), and healthy uninfected controls (Control).

2. Materials and Methods

2.1. Study Subjects

A study group consisted of 97 adults that were vaccinated with *M. bovis* BCG in childhood, including 51 patients with active pulmonary ATB (40 males, 11 females) aged 23–80 years, hospitalized and diagnosed in the Regional Center Hospital for Tuberculosis, Lung Diseases, and Rehabilitation in Lodz, Poland, 24 healthy volunteers, eight males, 16 females, with LTB (25–65 years old), and 22 healthy uninfected Controls (five males, 17 females), aged 18–66 years (Table 1). Active pulmonary TB was diagnosed by chest radiography and standard clinical examination—by Ziehl–Neelsen staining of sputum smears and *M.tb* culture as a gold standard. The Ethics Committee of the University in Lodz, Poland approved the study (ethical approval number 17/KBBN-UŁ/II/2016; date 2016/11/10). Informed consent to use blood for research purposes was signed by all participants.

Table 1. Patient characteristics.

	ATB	LTB	Control
N	51	24	22
Sex M/F	40/11 *	8/16	5/17 *
Ethnicity	Caucasian	Caucasian	Caucasian
Age			
median	54	51	37
range	23–80	25–65	18–66
years (IQR)	42–63	45–57	27–42
BCG vaccination	100%	100%	100%
QFT result, N (%)			
positive	22 (43%)	24 (100%)	0 (0%)
negative	28 (57%)	0 (0%)	22 (100%)

Abbreviations: ATB—active tuberculosis patients; LTB—latently *M.tb* infected individuals; Control—*M.tb*-uninfected healthy controls; QFT—QuantiFERON TB Gold test. * The proportion of men in the ATB group was significantly higher than in the Control group ($p < 0.05$).

2.2. RNA Isolation

RNA isolation from the buffy coat obtained after centrifugation (150 g, 4 °C, 10 min.) was performed by the use of a commercial QIAamp® RNA Blood Mini set. Genetic material was isolated from 3.5 mL of peripheral blood obtained from all volunteers using EDTA tubes and the BD Vacutainer® Blood Collection system. The isolation process was fully compliant with the manufacturer's guidelines. The isolation process was extended by an additional purification step using the RNase-Free DNase Set to obtain the purest product free of any genomic DNA. All of the procedures were carried out within no more than 2 h from the collection of a blood sample. Part of extracted RNA was used to visualize a product and obtain cDNA immediately after the isolation process; the rest of the genetic material was stored at −80 °C until analyzed.

2.3. Spectrophotometric Evaluation of Isolated RNA and Gel Visualization

At the end of the isolation procedure, 1.5 µL of each sample was pipetted into a sterile Eppendorf tube, which was then placed in an ice block, in order to assess the quality and quantity of RNA obtained. An additional water-containing blank was prepared to calibrate the device (NanoDrop), which was used for RNA elution in the final isolation step. For visualization of RNA, 1.2% agarose gel was prepared based on TAE buffer. Agarose gel was enriched after cooling with 10 µL of 5 mg/mL ethidium bromide. Each sample of RNA in a volume of 3.5 µL was heated to 70 °C in a water bath for 1 min. and then cooled on ice for another minute. An equal amount of loading buffer was then added to each of the samples, and the mixture was then loaded to the gel. The GeneRuler Plus DNA Ladder size 100 bp from ThermoFisher was used as the size standard. Electrophoresis was carried out at 90 V for 60 min. Subsequently, to visualize the obtained product, the gel was transferred to a Gel-Doc 2000 apparatus that was connected to a computer with Quantity One software. The analysis of the obtained gel allowed a clear distinction between the two isolated RNA fractions: 18S RNA and 28S RNA.

2.4. Reverse Transcription

cDNA was synthesized according to the manufacturer's instructions of the iScript™ cDNA Synthesis Kit (Bio-Rad). In the first stage of the reverse transcription reaction, 1 μg matrix RNA previously tested for quality and integrity was transferred from an isolated sample to 0.2 μL Eppendorf tubes, followed by the addition of reagents that are necessary for the cDNA synthesis process according to the manufacturer's proportions, which results in a mixture with a final volume of 20 μL. The reverse transcription reaction was carried out in a Biometra UNO II thermal cycler under conditions following the manufacturer's guidelines. The resulting cDNA was stored at −20 °C until analyzed.

2.5. qPCR Reaction

qPCR was performed in a CFX96 Real-Time PCR Detection System (Bio-Rad). The reaction mixture (10 μL) contained 5 μL of iTaq universal SYBR Green Supermix, 0.5 μL of each primer (10 μM), 1 μL of cDNA, and 3 μL of nuclease-free water. Amplifications were performed using the following cycling profile: an initial activation step (95 °C for 3 min.) followed by 40 cycles of denaturation at 95 °C for 10 s, annealing at a temperature appropriate for selected starters (Table 2) for 10 s, and extension at 72 °C for 20 s. For melting curve analysis, a dissociation step cycle (60 °C for 5 s, and then 0.5 °C for 5 s until 95 °C) was added. All of the qRT-PCR experiments were performed in three technical replicas.

Table 2. Starters and temperatures of annealing selected for expression analysis.

	Sequence	Temperature of Annealing	Source
IL-18	forward 5′-GCTTGAATCTAAATTATCAGTC-3′ reverse 5′-GAAGATTCAAATTGCATCTTAT-3′	55 °C	[22]
IL-18BP	forward 5′-CAACTGGACACCAGACCTCA-3′ reverse 5′-AGCTCAGCGTTCCATTCAGT-3′	64 °C	[23]
IL-18R	forward 5′-GGACTCCATGAAGCATTGGT-3′ reverse 5′-AGACTCGGAAAGAACAGGCA-3′	58 °C	[24]
IFN-γ	forward 5′-CTCTTGGCTGTTACTGCCAGG-3′ reverse 5′-CTCCACACTCTTTTGGATGCT-3′	60 °C	[25]
IL-37	Sino Biological INC.	60 °C	-

Analysis of gene expression was done through a comparative method (ΔΔCt) in order to determine the relative level of expression of selected mRNAs. This method is based on calculating the differences in the level of expression of the test gene and the reference gene. The calculations use the threshold cycle (Ct) values of the qPCR reaction. Ct values were determined for both the test and reference genes in both the test and control samples, for which the differences between the individual Ct values (ΔCt) were then calculated.

2.6. Statistical Analysis

The expression of genes between the study groups was compared using Kruskal–Wallis' non-parametric diagnostic test. A p-value < 0.05 was considered to be statistically significant. Statistical analyses were done using MedCalc (MedCalc Software, Ostend, Belgium) and GraphPad Prism 8 (GraphPad Software, La Jolla, CA, USA) software.

3. Results

3.1. Selection of Reference Genes

We had chosen to use the Reference Gene qPCR Panel BioRad with primers being designed and coated on the plate in lyophilized form for selected reference genes adequate for the analyzed material obtained from volunteers [26]. The housekeeping genes were: ACTB, RPL13A, B2M, RPLP0, G6PD, RPS18, GAPDH, TBP, GUSB, TFRC, HMBS, YWHAZ, HPRT1, PGK1, and IPO8. Additionally, the panel contained several internal controls ensuring the most reliable results. After the RT-PCR reaction, the analysis of the melting curves for the obtained amplicons was carried out in order to determine the quality of the reaction. In the next stage, the Ct values that were detected for individual genes were used for further analyses, which were carried out using the BioRad program (CFX Manager™ Software), and the GeNorm program. As a consequence of the analysis, we decided to use the following reference genes: GADPH, HPRT1, and TBP. These genes had the lowest M value corresponding to the most stable gene expression in the sample tested. The analysis of Ct values using the geNorm program allowed for further refinement of this gene list and, in turn, only HRTP1 and GAPDH were used as the reference panel, as shown in Figure 1.

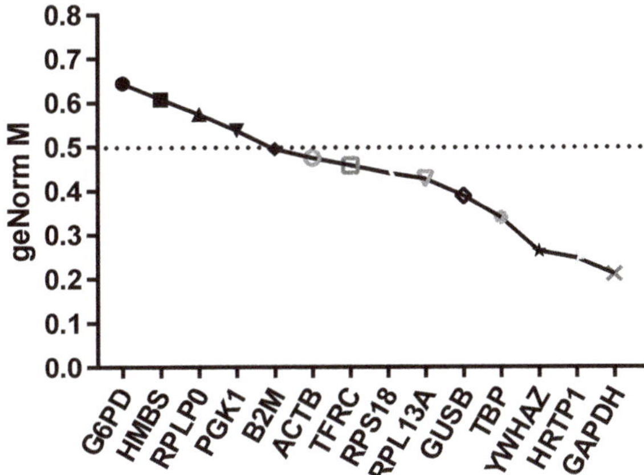

Figure 1. Chart showing the most (lowest M value) and least stable (highest M value) reference genes indicated by the program GeNorm.

Melt peak analysis demonstrated a single homogenous peak for all primer sets, including selected reference genes (Figure 2).

3.2. IL-18, IL-18BP, IL-18R, IFN-γ, and IL-37 Gene Expression in the Studied Groups

A significantly higher relative level of IL-18 mRNA expression was observed in LTB individuals as compared to healthy controls without *M.tb* infection ($p = 0.023$). A similar increase in the relative IL-18 expression level was observed among ATB patients; however, the difference in values for the group ATB and group Control group was not significant ($p = 0.082$) (Figure 3A).

The relative level of IL-18BP mRNA expression was significantly higher in ATB patients ($p < 0.001$) and healthy LTB individuals ($p = 0.006$) than in Control group volunteers (Figure 3B).

There were major significant differences between the three groups in the IL-18R mRNA expression. In the LTB groups, the relative level of IL-18R mRNA was much lower than in the ATB patients ($p < 0.001$) and individuals from the Control group ($p < 0.001$) (Figure 3C).

The level of relative expression of IFN-γ mRNA was significantly higher in active TB patients ($p = 0.002$) and LTB individuals ($p = 0.029$) than in the Control group (Figure 3D).

No statistically significant differences were observed in the levels of relative expression of IL-37 mRNA among the studied groups (Figure 3E).

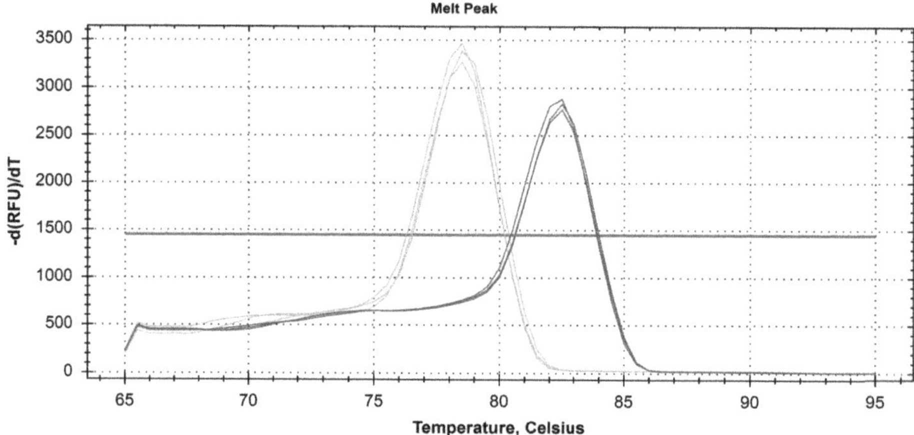

Figure 2. Melt peak analysis of reference genes HPRT1 (light grey) and GAPDH (dark grey).

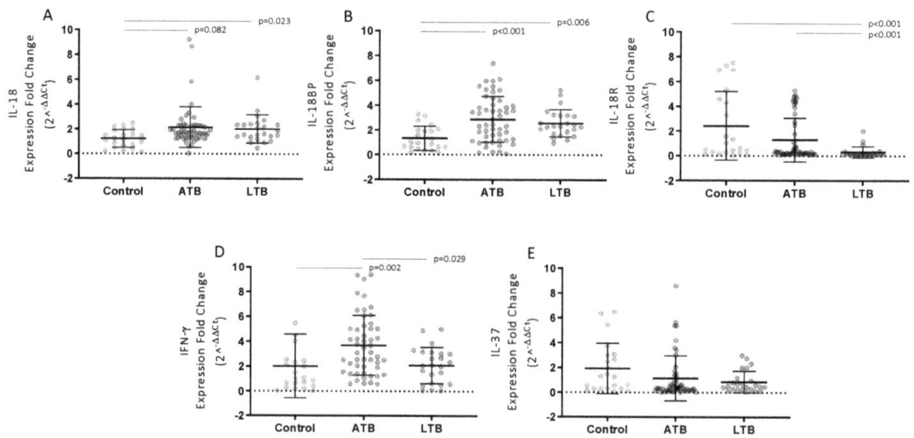

Figure 3. Relative expression of IL-18, IL-18BP, IL-18R, IFN-γ, and IL-37 mRNA in studied groups. Dot plot with mean (horizontal line), and standard deviation (whiskers), IL-18 (**A**), IL-18BP (**B**), IL-18R (**C**), IFN-γ (**D**), IL-37 (**E**) in the groups of healthy volunteers (Control), patients with active tuberculosis (ATB), and latently infected individuals (LTB).

4. Discussion

A unique feature of *M.tb* is the ability to persist in the host for a long time, despite functioning mechanisms of acquired immunity. In turn, approximately 2.3 billion individuals have latent tuberculosis infection without evidence of the clinical manifestation of active TB. It is hypothesized that active TB is usually caused by the reactivation of endogenous infection and untreated LTB is a major source of new active TB infections and transmission. The risk for severe active tuberculosis reactivation is increased several times among the immunocompromised individuals, diabetics, organ transplantation recipients, patients with hematologic malignancies, or HIV-infected subjects. Several treatment regimens of LTB are recommended, between isoniazid monotherapy for six months, rifampicin plus isoniazid for three months or rifapentine plus isoniazid for three months [27]. Currently, interferon-gamma release assays (IGRA) are used to test LTB based on the production of IFN-γ by Th1 cells responding to specific *M.tb* antigens, although the IGRA's accuracy in immunocompromised individuals is still limited. Moreover, the definition of active tuberculosis infection might not be accurate using IFN-γ responses to *M.tb* antigens even in combination with tuberculin skin test [28]. At the same time, neither the IGRA test nor the tuberculin skin test can distinguish between latent and active TB infection. Strategies for rapid differentiation of patients with active TB and people with LTB and prevention of tuberculosis reactivation in LTB individuals are urgently needed.

To meet such needs, we compared the expression of genes of the IL-18 pathway, functional receptor of this cytokine IL-18R, and IFN-γ, as well as the expression of IL-18BP and IL-37 genes in groups of patients with active TB, healthy individuals with LTB and healthy controls without *M.tb* infection.

Our results for the first time showed a significant increase in the relative expression of IL-18 and IL-18BP mRNA in the group of patients with active TB and LTB individuals when compared to healthy controls, according to available literature data. It might suggest a permanent activation of the immune cell signaling pathways in the course of *M.tb* infection either in the control or progression to active TB disease. Moreover, no significant differences in relative IL-18 and IL-18BP mRNA expression were observed between active TB patients and LTB individuals. Pechkovsky et al. demonstrated an increased expression of IL-18 mRNA in type II lung epithelial cells obtained from patients with pulmonary TB. Pneumocytes that were cultured in the presence of *M.tb* cell lysate showed an increased IL-18 mRNA expression in comparison with unstimulated cells and pneumocytes stimulated with PPD or LPS [29]. Higher levels of IL-18 mRNA expression were observed in monocytes responding to *M. leprae* antigens [30]. Corbaz et al. showed significantly higher levels of IL-18 and IL-18BP mRNA expression in the intestinal mucosa of patients with Crohn's disease as compared to healthy controls. We detected a similar increase in IL-18 and IL-18BP mRNA expression in the group of patients with active TB and healthy LTB individuals. These results are somewhat out of line with our earlier definition of a simultaneous increase in serum IL-18 and IL-18BP protein expression, which might be treated as a discriminatory biomarker of active tuberculosis and LTB. Yet, it is worth noting that only the complex co-expression of serum IL-18BP and IL-37, IP-10, and IFN-γ were identified as the most accurate discriminative biomarker set for diagnosis of active TB [21].

Our study allowed for the discovery of a novel relation—the significantly lower expression of functional IL-18R receptor mRNA in the LTBI group as compared to the active TB and healthy controls. This low expression of IL-18R mRNA in LTB individuals was accompanied by IFN-γ mRNA expression at the 'baseline' level, characterizing healthy individuals without *M.tb* infection. Similarly, Taha et al. pointed out that the expression of IFN-γ genes was significantly higher in the group of patients with active TB as compared to those who were infected but did not develop active TB [31].

Among many possible mediators of host response to *M.tb* is the activity of the indoleamine 2,3-dioxygenase-1 (IDO1), the enzyme of tryptophan metabolism, which leads to the formation of tryptophan metabolites, including quinolinic and picolinic acids [32]. In animal models, increased IDO-1 expression and the activation of the tryptophan-kynurenine pathway were indicated to play a crucial role in *M.tb* pathogenesis [29,30,33,34]. The depletion of tryptophan, which is required for microbial growth, as well as the accumulation of biologically active tryptophan metabolites, impaired effective anti-mycobacterial immune response and, thus, favoured survival and persistence of the pathogen [32]. In *M.tb*-infected mice, IFN-γ receptor-deficiency in nonhaematopoietic cells led to a lack of IDO-1 expression and it was associated with exuberant neutrophil recruitment and increased mortality [33]. In macaques, the suppression of IDO activity led to the reduction of the bacterial burden and clinical symptoms of active TB that was accompanied by increased lung T cell proliferation, the induction of inducible bronchus-associated lymphoid tissue, and the relocation of effector T cells to the center of the granuloma [34]. Mehra et al. demonstrated that IDO induction in the periphery of the granuloma correlated with active TB disease [30]. A few studies have investigated IDO-mediated tryptophan metabolism and its metabolites in humans in the context TB. Li et al. demonstrated increased IDO expression and activity in the pleural fluid from TB patients [35]. Almeida et al. found significantly higher expression levels of immune-suppressive mediators, including IDO-1, in patients with active pulmonary TB as compared to patients with other infectious lung diseases and healthy volunteers [36]. The authors suggested that the increased levels of immunosuppressive mediators may render the immune activation and counteract the development of Th1-type immune response against *M.tb*. The IDO levels were elevated at the time of TB diagnosis and declined after TB treatment, which serves as evidence that IDO expression might be both: a useful diagnostic marker of active TB as well as prognostic factor in TB treatment of HIV-negative patients [36]. Additionally, Adu-Gyamfi et al. showed that plasma IDO expression is a potential biomarker of active TB in HIV-positive patients [37], while Shi et al. confirmed that IDO activity might have an auxiliary diagnosis value for the early discrimination of multi-drug resistant TB patients [38]. The increase in IDO activity was noticed in both HIV-infected and uninfected active TB patients as compared with individuals with latent TB infection [36,37,39]. In relation to this statement, the simultaneous reduction of IL-18R mRNA expression together with significant overexpression of IL-18 mRNA observed by us in the LTB group is of particular interest.

The presented results entitle us to hypothesize that: the increase in IL-18 gene expression, the lack of increase in IFN-γ gene expression, and the remarkably reduced expression of IL-18R gene may be a novel set of conditions that partially describe the homeostasis between *M.tb* and host-immunity in latent tuberculosis infection. As an obligate intracellular pathogen, *M.tb* has numerous adaptive mechanisms of modifying cellular processes in the fight against the host immune response. In latent TB infection, *M.tb* bacilli benefit from epigenetic changes that occurred in the host immune system under mycobacterial infection [40]. These changes make the *M.tb* favorable environment in the host cells and promote mycobacterial survival, growth, and latency. In a study that was conducted among Chinese patients with pulmonary TB and healthy controls, single nucleotide polymorphisms in the IL-18R promoter were associated with genotype-specific methylation status and genotype-specific IL-18R expression [41]. In the author's opinion, the relationship between decreased mRNA expression of IL-18R that is caused by an SNP and increased DNA methylation can partially mediate the susceptibility to TB risk. No statistically significant differences were observed in the relative mRNA IL-37 expression among the groups in our study. IL-37 is a new member of the IL-1 family, which reduces systemic and local inflammation. IL-37 is expressed in various cells and tissues and it is regulated by numerous inflammatory stimuli and cytokines via different signal transduction pathways [42]. Mannose-capped lipoarabinomannan purified from *M.tb* induces IL-37 production via enhancing TLR2 expression in human type II alveolar epithelial cells; this process might contribute to the persistence of *M.tb* infection [43]. Zhao et al. indicated that IL-37 is also a negative regulator of immune responses in *Listeria monocytogenes* infection due to reduced production of colony-stimulating factors and increased macrophage apoptosis [44]. In our earlier studies, the serum concentration of the IL-37 protein was similar in the group of patients with active pulmonary TB and healthy individuals with or without latent *M.tb* infection [21]. However, the complex co-expression between the two IL-18 inhibitors, IL-18BP and IL-37, was identified as the strongest discriminative biomarker of active TB disease.

5. Conclusions

The role of IL-18, its binding protein IL-18BP, and IFN-γ in the development of the immune response against mycobacteria was confirmed by observing the increased level of the expression of these genes in the group of patients with active pulmonary TB. The reduced expression of IL-18R gene in healthy individuals with latent TB infection can, at least partially, prevent the development of a pathological inflammatory reaction and promote the maintenance of homeostatic conditions between host immunity and *M.tb* infection. In our future studies, we plan to test the expression of other genes that are tightly co-regulated with the IL-18 pathway.

Author Contributions: Conceptualization: S.W., M.D.; Data curation: S.W., G.K.; Formal analysis: M.D.; Funding acquisition: S.W., M.D.; Investigation: S.W.; Methodology: S.W., M.S.; Supervision: G.K., W.R., M.D.; Validation: S.W.; Visualization: S.W.; Writing original draft: S.W., W.R., M.S., M.D. All authors have read and agreed to the published version of the manuscript.

Funding: This work was supported by the National Science Centre grants no 2015/19/N/NZ6/01385 and 2016/21/B/NZ7/01771.

Conflicts of Interest: The authors declare that there is no conflict of interest regarding the publication of this article.

References

1. World Health Organisation. *Global Tuberculosis Report 2018*; World Health Organization: Geneva, Switzerland, 2018. Available online: https://apps.who.int/iris/handle/10665/274453 (accessed on 5 June 2020).
2. Ahmad, S. Pathogenesis, immunology, and diagnosis of latent mycobacterium tuberculosis infection. *Clin. Dev. Immunol.* **2011**, *2011*, 814943. [CrossRef] [PubMed]
3. Akdis, M.; Aab, A.; Altunbulakli, C.; Azkur, K.; Costa, R.A.; Crameri, R.; Duan, S.; Eiwegger, T.; Eljaszewicz, A.; Ferstl, R.; et al. Interleukins (from IL-1 to IL-38), interferons, transforming growth factor β, and TNF-α: Receptors, functions, and roles in diseases. *J. Allergy Clin. Immunol.* **2016**, *138*, 984–1010. [CrossRef] [PubMed]
4. Wawrocki, S.; Druszczynska, M.; Kowalewicz-Kulbat, M.; Rudnicka, W. Interleukin 18 (IL-18) as a target for immune intervention. *Acta Biochim. Pol.* **2016**, *63*, 59–63. [CrossRef] [PubMed]
5. Schneider, B.E.; Korbel, D.; Hagens, K.; Koch, M.; Raupach, B.; Enders, J.; Kaufmann, S.H.E.; Mittrücker, H.-W.; Schaible, U.E. A role for IL-18 in protective immunity against *Mycobacterium tuberculosis*. *Eur. J. Immunol.* **2010**, *40*, 396–405. [CrossRef] [PubMed]
6. Nakanishi, K.; Yoshimoto, T.; Tsutsui, H.; Okamura, H. Interleukin-18 regulates both Th1 and Th2 response. *Annu. Rev. Immunol.* **2001**, *19*, 423–474. [CrossRef] [PubMed]
7. Wawrocki, S.; Kielnierowski, G.; Rudnicka, W.; Druszczynska, M. Lack of significant effect of interleukin-18 gene variants on tuberculosis susceptibility in the Polish population. *Acta Biochim. Pol.* **2019**, *66*, 337–342. [CrossRef] [PubMed]
8. Novick, D.; Kim, S.; Kaplanski, G.; Dinarello, C.A. Interleukin-18, more than a Th1 cytokine. *Semin. Immunol.* **2013**, *25*, 439–448. [CrossRef]
9. Dinarello, C.A.; Novick, D.; Kim, S.; Kaplanski, G. Interleukin-18 and IL-18 binding protein. *Front. Immunol.* **2013**, *4*, 1–10. [CrossRef]
10. Gracie, J.A.; Robertson, S.E.; Mcinnes, I.B. Interleukin-18 Abstract. *J. Leukoc. Biol.* **2003**, *73*, 213–224. [CrossRef]
11. Huang, Y.; Li, D.; Zhang, P.; Liu, M.; Liang, X.; Yang, X.; Jiang, L.; Zhang, L.; Zhou, W.; Su, J.; et al. IL-18R-dependent and independent pathways account for IL-18-enhanced antitumor ability of CAR-T cells. *FASEB J.* **2020**, *34*, 1768–1782. [CrossRef]
12. Gutzmer, R.; Langer, K.; Mommert, S.; Wittmann, M.; Kapp, A.; Werfel, T. Human Dendritic Cells Express the IL-18R and Are Chemoattracted to IL-18. *J. Immunol.* **2003**, *171*, 6363–6371. [CrossRef] [PubMed]
13. Tiret, L.; Godefroy, T.; Lubos, E.; Nicaud, V.; Tregouet, D.-A.; Barbaux, S.; Schnabel, R.; Bickel, C.; Espinola-Klein, C.; Poirier, O.; et al. Genetic analysis of the interleukin-18 system highlights the role of the interleukin-18 gene in cardiovascular disease. *Circulation* **2005**, *112*, 643–650. [CrossRef] [PubMed]
14. Nold, M.F.; Nold-Petry, C.A.; Zepp, J.A.; Palmer, B.E.; Bufler, P.; Dinarello, C.A. IL-37 is a fundamental inhibitor of innate immunity. *Nat. Immunol.* **2010**, *11*, 1014–1022. [CrossRef] [PubMed]
15. Dinarello, C.A.; Nold-Petry, C.; Nold, M.; Fujita, M.; Li, S.; Kim, S.; Bufler, P. Suppression of innate inflammation and immunity by interleukin-37. *Eur. J. Immunol.* **2016**, *46*, 1067–1081. [CrossRef]
16. Quirk, S.; Agrawal, D.K. Immunobiology of IL-37: Mechanism of action and clinical perspectives. *Expert Rev. Clin. Immunol.* **2014**, *10*, 1703–1709. [CrossRef]
17. Pan, Y.; Wen, X.; Hao, D.; Wang, Y.; Wang, L.; He, G.; Jiang, X. The role of IL-37 in skin and connective tissue diseases. *Biomed. Pharmacother.* **2020**, *122*, 109705. [CrossRef]

18. Donovan, M.L.; Schultz, T.E.; Duke, T.J.; Blumenthal, A. Type I interferons in the pathogenesis of tuberculosis: Molecular drivers and immunological consequences. *Front. Immunol.* **2017**, *8*, 1633. [CrossRef]
19. Naylor, S.L.; Sakaguchi, A.Y.; Shows, T.B.; Law, M.L.; Goeddel, D.V.; Gray, P.W. Human immune interferon gene is located on chromosome 12. *J. Exp. Med.* **1983**, *157*, 1020–1027. [CrossRef]
20. Druszczynska, M.; Wlodarczyk, M.; Kielnierowski, G.; Seweryn, M.; Wawrocki, S.; Rudnicka, W. CD14-159C/T polymorphism in the development of delayed skin hypersensitivity to tuberculin. *PLoS ONE* **2017**, *12*, e0190106. [CrossRef]
21. Wawrocki, S.; Seweryn, M.; Kielnierowski, G.; Rudnicka, W.; Wlodarczyk, M.; Druszczynska, M. IL-18/IL-37/IP-10 signalling complex as a potential biomarker for discriminating active and latent TB. *PLoS ONE* **2019**, *14*, e0225556. [CrossRef]
22. Wang, Y.; Shan, N.; Chen, O.; Gao, Y.; Zou, X.; Wei, D.; Wang, C.; Zhang, Y. Imbalance of Interleukin-18 and Interleukin-18 Binding Protein in Children with Henoch-Schönlein Purpura. *J. Int. Med. Res.* **2011**, *39*, 2201–2208. [CrossRef] [PubMed]
23. Reale, M.; Kamal, M.A.; Patruno, A.; Costantini, E.; D'Angelo, C.; Pesce, M.; Greig, N.H. Neuronal cellular responses to extremely low frequency electromagnetic field exposure: Implications regarding oxidative stress and neurodegeneration. *PLoS ONE* **2014**, *9*, 1–14. [CrossRef] [PubMed]
24. Yoshino, O.; Osuga, Y.; Koga, K.; Tsutsumi, O.; Yano, T.; Fujii, T.; Kugu, K.; Momoeda, M.; Fujiwara, T.; Tomita, K.; et al. Evidence for the expression of interleukin (IL)-18, IL-18 receptor and IL-18 binding protein in the human endometrium. *MHR Basic Sci. Reprod. Med.* **2001**, *7*, 649–654. [CrossRef] [PubMed]
25. Zumwalt, T.J.; Arnold, M.; Goel, A.; Boland, C.R. Active secretion of CXCL10 and CCL5 from colorectal cancer microenvironments associates with GranzymeB+ CD8+ T-cell infiltration. *Oncotarget* **2015**, *6*, 2981–2991. [CrossRef]
26. Bustin, S.A.; Benes, V.; Garson, J.A.; Hellemans, J.; Huggett, J.; Kubista, M.; Mueller, R.; Nolan, T.; Pfaffl, M.W.; Shipley, G.L.; et al. The MIQE guidelines: Minimum information for publication of quantitative real-time PCR experiments. *Clin. Chem.* **2009**, *55*, 611–622. [CrossRef]
27. World Health Organisation. *Latent Tuberculosis Infection Updated and Consolidated Guidelines for Programmatic Management*; World Health Organization: Geneva, Switzerland, 2018. Available online: https://www.who.int/tb/publications/2018/latent-tuberculosis-infection/en/ (accessed on 5 June 2020).
28. Wlodarczyk, M.; Rudnicka, W.; Janiszewska-Drobinska, B.; Kielnierowski, G.; Kowalewicz-Kulbat, M.; Fol, M.; Druszczynska, M. Interferon-gamma assay in combination with tuberculin skin test are insufficient for the diagnosis of culture-negative pulmonary tuberculosis. *PLoS ONE* **2014**, *9*, 1–12. [CrossRef]
29. Moreau, M.; Lestage, J.; Verrier, D.; Mormède, C.; Kelley, K.W.; Dantzer, R.; Castanon, N. Bacille Calmette-Guérin inoculation induces chronic activation of peripheral and brain indoleamine 2,3-dioxygenase in mice. *J. Infect. Dis.* **2005**, *192*, 537–544. [CrossRef]
30. Mehra, S.; Alvarez, X.; Didier, P.J.; Doyle, L.A.; Blanchard, J.L.; Lackner, A.A.; Kaushal, D. Granuloma correlates of protection against tuberculosis and mechanisms of immune modulation by Mycobacterium tuberculosis. *J. Infect. Dis.* **2013**, *207*, 1115–1127. [CrossRef]
31. Taha, R.A.; Kotsimbos, T.C.; Song, Y.L.; Menzies, D.; Hamid, Q. IFN-γ and IL-12 are increased in active compared with inactive tuberculosis. *Am. J. Respir. Crit. Care Med.* **1997**, *155*, 1135–1139. [CrossRef]
32. Blumenthal, A.; Nagalingam, G.; Huch, J.H.; Walker, L.; Guillemin, G.J.; Smythe, G.A.; Ehrt, S.; Britton, W.J.; Saunders, B.M.M. tuberculosis induces potent activation of IDO-1, but this is not essential for the immunological control of infection. *PLoS ONE* **2012**, *7*, 1–14. [CrossRef]
33. Desvignes, L.; Ernst, J.D. Interferon-γ-Responsive nonhematopoietic cells regulate the immune response to Mycobacterium tuberculosis. *Immunity* **2009**, *31*, 974–985. [CrossRef] [PubMed]
34. Gautam, U.S.; Foreman, T.W.; Bucsan, A.N.; Veatch, A.V.; Alvarez, X.; Adekambi, T.; Golden, N.A.; Gentry, K.M.; Doyle-Meyers, L.A.; Russell-Lodrigue, K.E.; et al. In vivo inhibition of tryptophan catabolism reorganizes the tuberculoma and augments immune-mediated control of Mycobacterium tuberculosis. *Proc. Natl. Acad. Sci. USA* **2018**, *115*, E62–E71. [CrossRef] [PubMed]
35. Li, Q.; Li, L.; Liu, Y.; Fu, X.; Qiao, D.; Wang, H.; Lao, S.; Huang, F.; Wu, C. Pleural fluid from tuberculous pleurisy inhibits the functions of T cells and the differentiation of Th1 cells via immunosuppressive factors. *Cell. Mol. Immunol.* **2011**, *8*, 172–180. [CrossRef] [PubMed]

36. Almeida, A.S.; Lago, P.M.; Boechat, N.; Huard, R.C.; Lazzarini, L.C.O.; Santos, A.R.; Nociari, M.; Zhu, H.; Perez-Sweeney, B.M.; Bang, H.; et al. Tuberculosis is associated with a down-modulatory lung immune response that impairs th1-type immunity. *J. Immunol.* **2009**, *183*, 718–731. [CrossRef]
37. Adu-Gyamfi, C.G.; Snyman, T.; Hoffmann, C.J.; Martinson, N.A.; Chaisson, R.E.; George, J.A.; Suchard, M.S. Plasma indoleamine 2, 3-dioxygenase, a biomarker for tuberculosis in human immunodeficiency virus-infected patients. *Clin. Infect. Dis.* **2017**, *65*, 1356–1358. [CrossRef]
38. Shi, W.; Wu, J.; Tan, Q.; Hu, C.M.; Zhang, X.; Pan, H.Q.; Yang, Z.; He, M.Y.; Yu, M.; Zhang, B.; et al. Plasma indoleamine 2,3-dioxygenase activity as a potential biomarker for early diagnosis of multidrug-resistant tuberculosis in tuberculosis patients. *Infect. Drug Resist.* **2019**, *12*, 1265–1276. [CrossRef]
39. Suzuki, Y.; Suda, T.; Asada, K.; Miwa, S.; Suzuki, M.; Fujie, M.; Furuhashi, K.; Nakamura, Y.; Inui, N.; Shirai, T.; et al. Serum indoleamine 2,3-dioxygenase activity predicts prognosis of pulmonary tuberculosis. *Clin. Vaccine Immunol.* **2012**, *19*, 436–442. [CrossRef]
40. Yadav, V.; Dwivedi, V.P.; Bhattacharya, D.; Mittal, A.; Moodley, P.; Das, G. Genetics and genome research understanding the host epigenetics in Mycobacterium tuberculosis infection. *J. Genet Genome Res.* **2015**, *2*, 1. [CrossRef]
41. Zhang, J.; Zheng, L.; Zhu, D.; An, H.; Yang, Y.; Liang, Y.; Zhao, W.; Ding, W.; Wu, X. Polymorphisms in the interleukin 18 receptor 1 gene and tuberculosis susceptibility among Chinese. *PLoS ONE* **2014**, *9*, e110734. [CrossRef]
42. Wang, L.; Quan, Y.; Yue, Y.; Heng, X.; Che, F. Interleukin-37: A crucial cytokine with multiple roles in disease and potentially clinical therapy (Review). *Oncol. Lett.* **2018**, *15*, 4711–4719. [CrossRef]
43. Huang, Z.; Zhao, G.W.; Gao, C.H.; Chi, X.W.; Zeng, T.; Hu, Y.W.; Zheng, L.; Wang, Q. Mannose-capped lipoarabinomannan from mycobacterium tuberculosis induces IL-37 production via upregulating ERK1/2 and p38 in human type ii alveolar epithelial cells. *Int. J. Clin. Exp. Med.* **2015**, *8*, 7279–7287. [PubMed]
44. Zhao, M.; Hu, Y.; Shou, J.; Su, S.B.; Yang, J.; Yang, T. IL-37 impairs host resistance to Listeria infection by suppressing macrophage function. *Biochem. Biophys. Res. Commun.* **2017**, *485*, 563–568. [CrossRef] [PubMed]

Article

Preclinical Evidence of Nanomedicine Formulation to Target *Mycobacterium tuberculosis* at Its Bone Marrow Niche

Jaishree Garhyan [1], Surender Mohan [1], Vinoth Rajendran [2,3] and Rakesh Bhatnagar [1,4,*]

1 Laboratory of Molecular Biology and Genetic Engineering, School of Biotechnology, Jawaharlal Nehru University, New Delhi 110067, India; Jgarhyan@gmail.com (J.G.); mohan.surender@gmail.com (S.M.)
2 Department of Biochemistry, University of Delhi South Campus, Benito Juarez Road, New Delhi 110021, India; vinoth.avj@gmail.com
3 Department of Microbiology, School of Life Sciences, Pondicherry University, Puducherry 605014, India
4 Banaras Hindu University, Varanasi, Uttar Pradesh 221005, India
* Correspondence: vc@bhu.ac.in or rakeshbhatnagar@jnu.ac.in; Tel.: +91-011-26704079

Received: 22 March 2020; Accepted: 26 April 2020; Published: 13 May 2020

Abstract: One-third of the world's population is estimated to be latently infected with *Mycobacterium tuberculosis* (Mtb). Recently, we found that dormant Mtb hides in bone marrow mesenchymal stem cells (BM-MSCs) post-chemotherapy in mice model and in clinical subjects. It is known that residual Mtb post-chemotherapy may be responsible for increased relapse rates. However, strategies for Mtb clearance post-chemotherapy are lacking. In this study, we engineered and formulated novel bone-homing PEGylated liposome nanoparticles (BTL-NPs) which actively targeted the bone microenvironment leading to Mtb clearance. Targeting of BM-resident Mtb was carried out through bone-homing liposomes tagged with alendronate (Ald). BTL characterization using TEM and DLS showed that the size of bone-homing isoniazid (INH) and rifampicin (RIF) BTLs were 100 ± 16.3 nm and 84 ± 18.4 nm, respectively, with the encapsulation efficiency of 69.5% ± 4.2% and 70.6% ± 4.7%. Further characterization of BTLs, displayed by sustained in vitro release patterns, increased in vivo tissue uptake and enhanced internalization of BTLs in RAW cells and CD271+BM-MSCs. The efficacy of isoniazid (INH)- and rifampicin (RIF)-loaded BTLs were shown using a mice model where the relapse rate of the tuberculosis was decreased significantly in targeted versus non-targeted groups. Our findings suggest that BTLs may play an important role in developing a clinical strategy for the clearance of dormant Mtb post-chemotherapy in BM cells.

Keywords: BM-MSCs; Mtb; bone-homing; stem cell niche; latent tuberculosis; relapse; liposomes

1. Introduction

Tuberculosis (TB) remains one of the most common human diseases today, causing nearly two million deaths per year [1]. One-third of the global population is estimated to be latently infected with *Mycobacterium tuberculosis*, which is attributed to the ability of the tubercle bacillus to remain dormant in its protective niche unrecognized by the host immune system [1–6]. In post-chemotherapy patients, the recurrence of TB is very common due to endogenous reactivation leading to death [5,7–10]. It is imperative to find strategies that could effectively target and facilitate dormant *Mycobacterium tuberculosis* (Mtb) clearance to avert disease reactivation and associated mortality.

CD271+bone marrow mesenchymal stem cells (BM-MSCs) serve as a novel host and a protective niche for non-replicating dormant *Mycobacterium tuberculosis* [11–14]. Mycobacterial bacilli remains intracellularly protected by standard anti-Tb drug treatment in mesenchymal stem cell populations, as shown in murine models and post-chemotherapy TB patients [11,13–15]. Das et al. demonstrated that in post-DOT (direct observed treatment) patients, a specific population of MSCs with CD271+ surface marker, harbor live nonreplicating Mtb bacilli [12]. The residual dormant Mtb population post-drug treatment may lead to endogenous reactivation and develop resistant forms [5,6,10,16–18]. We and others have shown that in murine models, prolonged anti-Tb drugs at standard dosages fail to clear the Mtb population inside CD271+BM-MSCs [13,14,19]. More importantly, despite 90 days of anti-Tb drug treatment in mice, CD271+BM-MSC-resident Mtb reactivates and leads to relapse [14]. We hypothesized that a BM-directed approach where anti-Tb drugs are delivered specifically to the bone microenvironment, can target Mtb inside BM-MSCs and reduce the relapse rate.

We and others have shown that generic nanoparticle formulations impart protection and provide countermeasures against various infections [20–26]. Currently, to reduce off-target toxicity and enhance effective tissue treatment, organ-homing nanoparticles (NPs) are being used as drug-delivery vehicles [27–31]. Among other NPs, liposomes are the most versatile and are widely used in clinical settings because of their biocompatibility and controlled release profile [21,27,30,32]. Recent studies show that the bone microenvironment could specifically uptake surface-modified liposomes and PLGA NPs (poly (lactic-co-glycolic acid)) compared to other organs [33–35]. Although for TB infections, tissue-specific liposomes have been reported previously [22,36–41], a novel approach of bone homing liposomes has not been addressed yet. In this study, we employed a biphosphate molecule, alendronate (Ald) as target moiety and PEG to formulated bone-homing liposomes in order to deliver the standard anti-Tb drugs rifampicin (RIF) and isoniazid (INH) right to the bone microenvironment. We propose that drug encapsulated bone-homing PEGylated liposome (BTL) nanoparticles can clear the Mtb bacilli residing in CD271+BM-MSCs and reduce the relapse rate.

2. Results

2.1. Designing of Alendronate BTL-NPs and Their Characterization

Alendronate tagging of PEG-PE was carried out for formulating Alen-PEG-PE as illustrated (Supplementary Figure S1). Resulting Alen-PEG-PE was used for the synthesis of BTL liposomes using additional PE-PEG and PC (Figure 1A,B). INH, RIF and Coumarin6 (fluorescent marker C6) were separately encapsulated in BTLs to formulate three separate BTLs: INH BTL, RIF BTL and C6 BTL followed by characterization (Figure 2). C6 BTLs were used for bone affinity experiments (Figure 3), cellular uptake (Figure 4) and tissue uptake studies (Figure 5C). Transmission electron microscopy results showed that all liposomal formulations consist of unilamellar vesicles in the size range of 80 to 110 nm having a spherical morphology. Dynamic light scattering data showed that the mean diameter of the prepared formulations, C6 BTLs, INH BTLS and RIF BTLs, were 109 ± 12.6 nm, 100 ± 16.3 nm and 84 ± 18.4 nm in diameter, respectively. Zeta potential values were negative which facilitates the long circulation of liposomes (Figure 2A–C). Average drug encapsulation efficiency determined for C6 BTLs, INH-BTLs and RIF BTLs were 70% ± 3.7%, 69.5% ± 4.2% and 70.6% ± 4.7%, respectively (Table 1). Non-targeted liposomes exhibited similar size range and drug encapsulation efficiency.

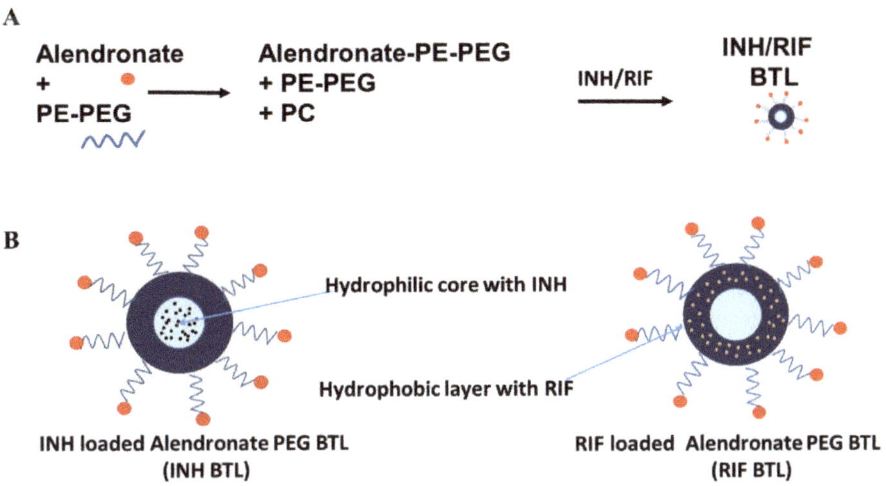

Figure 1. Designing of alendronate bone-homing PEGylated liposome (BTL) nanoparticles. (**A**) Alendronate and PE-PEG undergo chemical reactions for the binding of alendronate to form alendronate PE-PEG. Alendronate-PE-PEG is attached to the surface of liposomes. (**B**) Schematic of isoniazid (INH) and rifampicin (RIF) encapsulated alendronate PEG BTL-NPs. PC = Phosphatidyl-choline, $CH_2CH_2N^+(CH_3)_3$; PE = Phosphatidylethanolamine, $CH_2CH_2NH_3{}^+$.

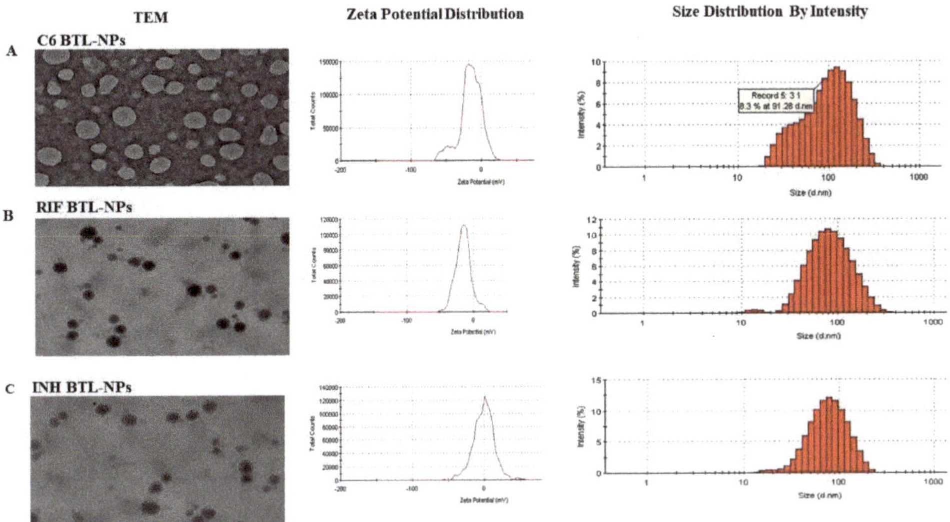

Figure 2. Characterization of liposomal formulations (TEM and size distribution of BTLs). Transmission electron microscopy (JEOL 2100F) images of all the three BTLs encapsulating C6, INH and RIF were performed. Dynamic light scattering and zeta potential were measured using Malvern Zetasizer. The above figure is a representative image for each liposomal formulation.

Table 1. Size and drug encapsulation efficiency of BTL-NPs.

Nanoparticle (NP)	NP Size (nm)	% Encapsulation
C6 BTL-NP	109 ± 12.6	70 ± 3.7
INH BTL-NP	100 ± 16.3	69.5 ± 4.2
RIF BTL-NP	84 ± 18.4	70.6 ± 4.7

2.2. Binding Assay of Alendronate BTL-NPs

Alendronate has a natural affinity toward the calcium component of Hydroxyapatite (HA) crystal. To investigate the affinity of engineered BTLs for BM, bone-chip binding and HA binding experiments were performed. In order to visualize qualitative binding, C6 (a fluorescent marker)-loaded BTLs were used. Alen-tagged BTLs showed a high fluorescence signal (green), whereas the non-targeted liposomes did not show any binding (Figure 3A), proving the high affinity of BTLs for mice bone fragments (femur). In the second binding assay, HA microparticles were allowed to bind with alen BTLs and non-targeted liposomes. TEM imaging results showed that HA binds to alen BTLs with greater affinity while non-targeted liposomes did not show any affinity to HA microparticles (Figure 3B). Overall, these results confirmed that BTLs exhibited a strong affinity to bone HA and therefore may possess bone-homing capabilities.

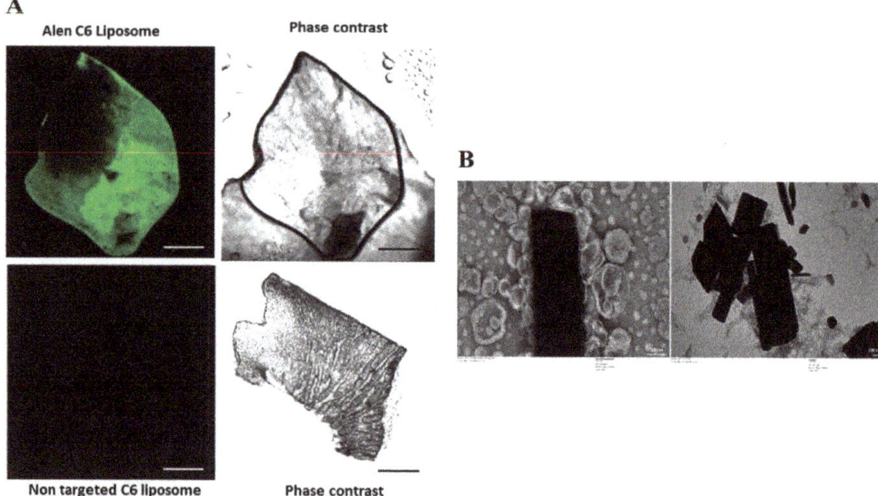

Figure 3. (**A**) Bone-chip binding assay of BTL-NPs. Bone fragments from femur were washed and incubated with alendronate BTLs encapsulating Coumarin6 (alen BTL C6) and non-targeted liposomes C6 (NT liposomes C6) for 30 min followed by washing and visualization through a confocal microscope (scale bar = 0.1 mm). (**B**) BTL's affinity to Hydroxyapatite (HA). For HA binding assay of BTL-NPs, HA microparticles were incubated with BTLs or NT liposomes. A 200 μL volume of 0.05 mg/mL HA was incubated with 40 μL of BTL or NT followed by rotation for 30 min at room temperature. The samples were washed 3 times gently and loaded on a TEM grid followed by staining and visualization with TEM (JEOL 2100F). Black rods are cylindrical HA microparticles and round nanoparticles around the HA are liposomes; scale bar = 100 nm HA + BTL (left panel), HA + NT (right panel).

2.3. Cellular Uptake of Alendronate BTL-NPs by CD271+BM-MSCs

To assess the cellular uptake efficiency of BTLs, two different cells were investigated: macrophages (RAW 264.7 cell line murine macrophages) and freshly isolated murine CD271+BM-MSCs. Using live cell imaging, RAW 264.7 cell uptake was observed from 5 min to 30 min after the addition of BTL. Figure 4A indicates the successful uptake of BTLs by macrophages. Next, mice CD271+BM-MSCs were incubated with alen BTLs which exhibited similar uptake patterns showing localization inside the cells at 30 min (Figure 4B).

A

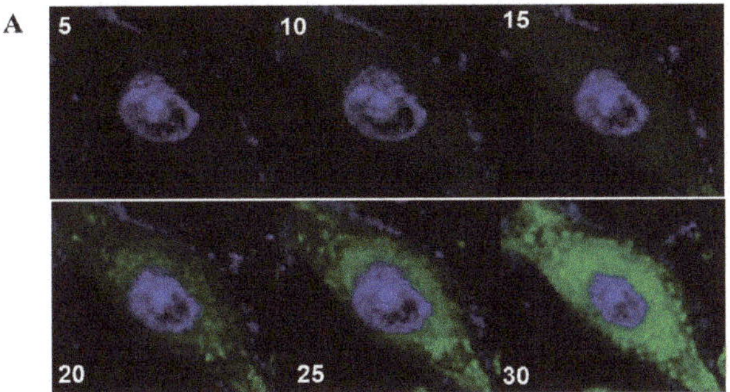

DAPI (Nucleus), C6BTL-NPs (targeted Nanoparticles)

B C6BTL-NPs | Phase contrast | Merged

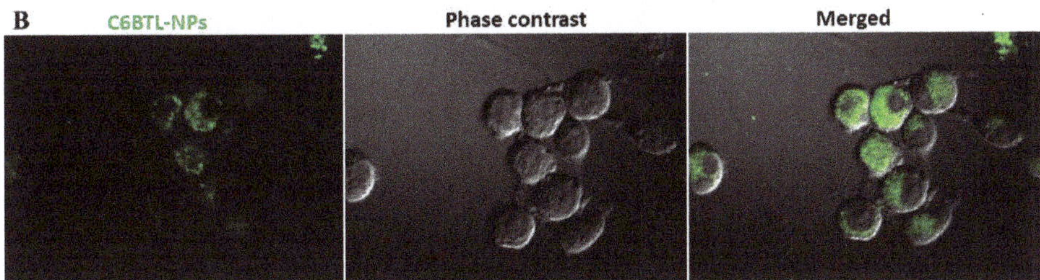

Figure 4. Confocal microscopy of C6BTLs uptake by CD271+BM-MSCs. Cells were incubated with C6BTL followed by live cell imaging to evaluate the time-bound uptake of BTLs by cells. (**A**) RAW 246.7 cells capture every 5 min up till 30 min after the addition of C6BTL. (**B**) CD271+BM-MSCs incubated with C6BTLs, fixed and imaged after 30 min.

2.4. In Vitro Drug Release Profile of INH and RIF BTL-NPs and Tissue Uptake

In vitro drug release profile was enumerated by an in vitro dialysis method. The INH-encapsulated alen BTLs and RIF-encapsulated alen BTLs showed sustained slower release till 72 h in comparison to their counterpart non-targeted NPs as indicated by % cumulative drug release pattern (CDR%). Notedly, INH-BTLs showed faster release compared to RIF BTLs (Figure 5A,B). Nevertheless, both targeted NP's cumulative drug release was slow and sustained. Further, we investigated the bone uptake of BTLs, where mice were injected with C6 loaded alen BTLs, C6 non-targeted (NT) liposomes or C6 PBS. Pairs of femur and tibia were excised followed by C6 extraction with chloroform: methanol. Fluorimetric analysis was carried out for determining the percentage of C6 dose injected, recovered from both bone pairs. We found that there was an approximate eight times increase in the homing of C6BTLs in comparison to C6NT liposomes (Figure 5C) demonstrating the bone-seeking nature of BTL-NPs.

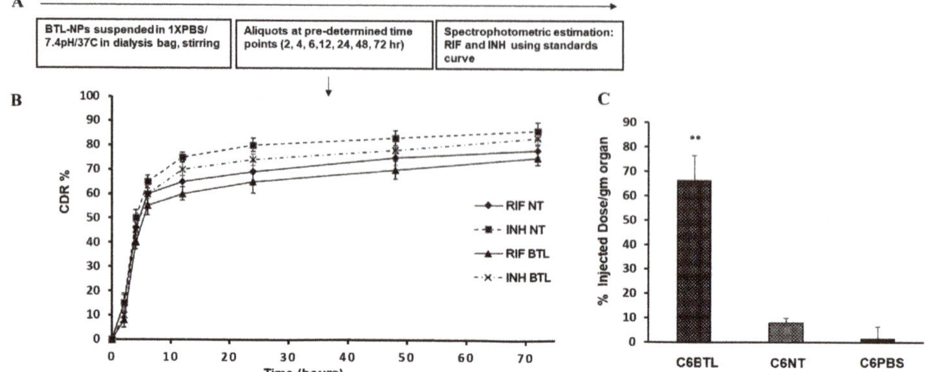

Figure 5. (**A**) Schematic of in vitro drug release. (**B**) Cumulative drug release pattern (CDR%) of targeted (BTLs) versus non-targeted (NT) for INH and RIF. A 500 μL volume of INH- and RIF-encapsulated BTLs and NTs were suspended in dialysis bags and stirred for 72 h in 25 mL of 1× PBS (pH 7.4). At each time point, 500 μL was aliquoted from a 25 mL reservoir for drug release assay. Data represents mean CDR% ± SEM (n = 3). (**C**) Tissue uptake of targeted (C6 BTL) versus non-targeted (C6 NT). Data are presented as % of C6 injected recovered per g of organ (organ = pair of tibia and femur). SEM (n = 3), ** $p < 0.001$ vs. C6NT.

2.5. *Efficacy of INH and RIF BTL-NPs in H37Rv Infected Mice*

For evaluating the efficacy of INH and RIF alendronate BTLs, standard Cornell model of persistent Mtb was followed (Figure 6A): Mice infected with pathogenic strains of Mtb H37RV were administered standard INH and RIF dosage for 90 days as a result of which all organs except for bone marrow were free of bacilli load [11,13]. After one week of the above regimen, animals were treated with the drug-loaded BTLs/NT NPs for 12 weeks, twice per week, with an effective concentration of INH = 4 mg/kg and RIF = 3 mg/kg. Appropriate controls were used. Animal health and weight were monitored for four months followed by CFU enumeration of CD271+BM-MSCs. A significant reduction of Mtb CFU in CD271+BM-MSCs was observed (Figure 6B). Notedly, Mtb culture from lungs was found negative in all experimental groups.

Relapse studies were carried out on Mtb-infected mice as illustrated in Figure 7A. Mice were monitored up to four months for relapse post-dexamethasone administration (10 mg/kg/ four weeks). Gross pathology inspection of the lung lobes was carried out for granuloma formation and the remaining portion of the lung was homogenized and cultured to score positive or negative Mtb load (Figure 7). Appropriate controls were included for relapse study viz. only vehicle and empty BTLs. All untreated mice succumbed to death within two months of infection. The relapse rate was 11% for the BTL group while the other experimental groups showed a higher relapse rate (Figure 7).

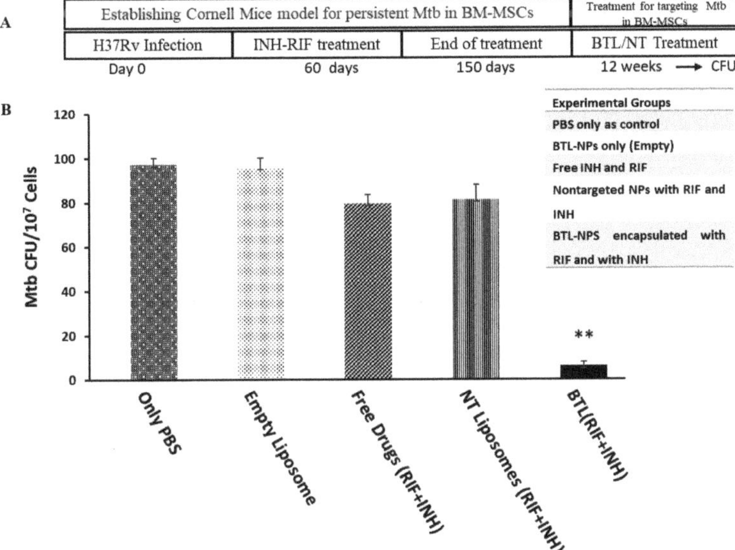

Figure 6. *Mycobacterium tuberculosis* (Mtb) load from CD271+BM-MSCs post-INH and RIF BTLs treatment. (**A**) Schematic of in vivo treatment. (**B**) C57bl/6 mice were infected with H37Rv, administered with standard anti-Tb drugs and next administered with drug-loaded BTL or NTs for 12 weeks/twice a week/intravenously. The effective dosage of INH and RIF was maintained at 4 mg/kg and 3 mg/kg, respectively, for all drug groups. Mtb CFU were enumerated by plating CD271+ BM-MSCs population obtained from immuno-magnetic sorting. $n = 3$, data expressed as mean CFU = +/− SEM per 10^7 cells. ** $p < 0.001$.

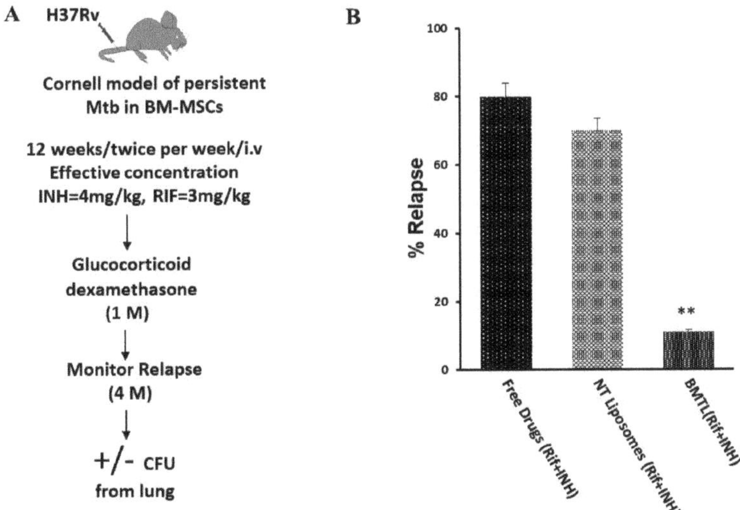

Figure 7. Relapse rate for INH and RIF BTLs versus NT liposomes. (**A**) Schematic of in vivo treatment for relapse. (**B**) Relapse % for targeted liposomes. Mice from each group were monitored for 4 months for signs of relapse. Animals were sacrificed for relapse signs: positive culture from lungs and gross lung pathology/granuloma formation. Data expressed as mean = +/− SEM. ** $p < 0.001$.

3. Discussion

We and others have shown that CD271+BM-Mesenchymal stem cells are natural reservoirs for dormant Mtb [11,13,14]. In this study, we engineered bone-homing liposomes decorated with Ald and PEG to actively target Mtb in its novel niche. Our BTL-NPs showed significant bone binding capabilities and resulted in the decreased Mtb load of CD271+BM-MSCs. Although bone-homing liposomes have been reported earlier [42,43], this is the first study to formulate and test alen-conjugated PEG liposomes to target BM-resident Mtb. Multiple functions were kept in mind and were incorporated into the engineering of BTL-NP: (1) bone-homing capability, (2) stealth properties by PEG and (3) size. For bone-homing, alendronate, a biphosphate was attached to the liposome surface resulting in high affinity and homing of BTLs to the bone microenvironment. Alendronate belongs to the bisphosphonate family of drugs and is generally used for cancer therapy. Previous studies show that alendronate has no cytotoxic effects and it does not alter the differentiation or self-renewal properties of stem cells residing in the bone marrow [35,43–45]. It is known that alendronate treatment can be helpful in the treatment of bone disease, especially bone cancer [46]. Alendronate, 5 or 10 mg/d, may increase the bone mineral density in women (postmenopausal) and men with primary osteoporosis [47]. Alendronic acid, administered at a 70 mg once weekly or 35 mg twice weekly dose, has shown to increase BMD (bone mineral density) as 10 mg/d and is already used clinically in the treatment or prevention of osteoporosis [48]. While alendronate on systemic administration homes to the bone microenvironment, PEG improves the systemic circulation of liposomes [34,49]. High circulation time owing to PEGylation, high bone-homing capability of BTL-NPs due to alendronate and optimal size in our study may have rendered them effective for clearance of CD271+BM-MSCs resident Mtb.

Our in vivo data indicates that BTL-NPs may have increased the available intracellular RIF and INH leading to Mtb clearance inside BM-MSCs. We also speculated that the PEG-modification of BTLs passively increased the bone marrow selectivity by inhibiting the hepatic uptake, resulting in BTL-NPs to target BM macrophages [35,50]. Moreover, it is possible that after getting trapped in macrophages, BTL-NPs gradually get digested by lysosomal enzymes leading to controlled drug release from macrophages to the surrounding BM tissue, further increasing intracellular drug. While MSCs are known to have drug efflux pumps that gradually efflux out drugs/antibiotics to protect the cells from cytotoxicity, our data suggest that the BTL-NPs may be delivering the drugs at a faster pace in comparison to their efflux rate. In aggregate, it is possible that BTL-NPs may increase the available intracellular drugs, rendering them more effective in Mtb killing compared [51] to free drugs or NT NPs. However, we acknowledge that this speculation requires further investigation.

Lung-homing liposomes targeting Mtb lowers treatment duration and hepatotoxicity in pulmonary tuberculosis [40,52,53]. Swami et al. recently showed that PLGA nanocarriers could be used for the targeted delivery of drugs to bone cancer [35]. These two studies taken together led to the hypothesis that BM-homing liposomes could aid in clearing tubercle bacilli present in the BM cells with reduced off-target toxicity. We acknowledge that, although the entire Mtb population from BM cells was not cleared leading to relapse in very few mice, a decrease in relapse rate is of considerable clinical significance. Few Mtb cells remaining in CD271+BM-MSC (Figure 6B) despite BTL treatment, indicate that there might be some subpopulation of Mtb which (a) may be pushed into deeper stage dormancy, (b) may have developed into drug resistance variants or (c) were present in an unreachable compartment of CD271+BM-MSCs. Any of these three possibilities require further investigation as deeper understanding may lead to complete clearance in BM cells. Nevertheless, currently, there is no treatment/drug which targets BM-MSCs resident Mtb and since alendronate is already in clinical use for bone strengthening, alendronate-based bone targeting may be of clinical significance. Once clinically investigated, BTLs may be used to reduce the relapse rate in post-chemotherapy patients. Overall, our study demonstrates that BTL-NPs can be used for the effective delivery of existing anti-TB drugs to bone microenvironment through specially formulated liposome carriers.

4. Materials and Methods

4.1. Alendronate Tagging of DSPE-PEG

First, 23 mg of DSPE-PEG (2000)-carboxylic acid (1,2-distearoyl-*sn*-glycero-3-phosphoethanolamine -N-[carboxy(polyethylene glycol)-2000]ammonium salt(Avanti lipids polar, Inc., USA, Cat Number 880125P) was dissolved in 5 mL of acetone. Next, 4.2 mg N,N-Dicyclohexylcarbodiimide (DCC) and 2.4 mg N-hydroxy succinimide (NHS) were added for activation overnight at RT. Syringe assisted removal of insoluble by-product (Dicyclohexylurea) was carried out. Following this drying of lipids was carried out for 2 h through nitrogen. For alendronate tagging, activated lipid and 2 mg alendronate sodium trihydrate (Ald, A-4978, Sigma-Aldrich, St. Louis, MO, USA) were dissolved in a mixture of DMSO and water for 24 h. This was followed by 24 h dialysis against water and subsequent drying under nitrogen.

4.2. Preparation of PEG Liposome and BTL PEGylated Liposome NPs with Ald Tagged PEG-PE and PC (with INH and RIF Encapsulation)

Liposome NPs were prepared as described previously [21,54,55]. In brief, Egg-PC and Egg-PE in a molar ratio of 8:2 were mixed in a 100 mL round bottom flask in chloroform and rotated under vacuum at 37 °C until a thin lipid layer is formed followed by desiccation for 2 h. For alendronate PEGylated liposomes a.k.a. BTLs, PC (2% of total lipids) + Ald-PE-PEG2000 (2.5% of total lipid content) + PE-mPEG2000 (2.5% of total lipid content) ratio was used. The resultant dry lipid was vortexed with PBS until complete lipid dispersion. For RIF-encapsulated BTL-NPs, Rifampicin (M.P Biomedicals, CAS #13294-4-1, Irvin, CA, USA) was dissolved in chloroform, and for INH-encapsulated BTL-NPs, isoniazid (Sigma-Aldrich, CAS #54-85-3, St. Louis, MO USA) was dissolved in saline at appropriate concentrations. Next, liposomal dispersion was centrifuged at 50,000 rpm (60 min) at 4 °C. This step was done to get rid of un-encapsulated drugs. The pellet obtained was resuspended and centrifuged twice to remove the traces of un-encapsulated drugs. For equal size and decreased variation, the final suspension obtained after three washing was extruded sequentially through 3 membrane filters. The liposomal suspension was assayed spectrophotometrically for the presence of RIF and INH, according to the modified method described [56]. Millipore membrane filters used were of Type RA from Millipore Corp., Bedford, MA, USA.

4.3. TEM and Size Distribution

The liposomal suspension was diluted in distilled water and placed upon 300-mesh carbon-coated copper grids and air dried for analysis [21,54]. The liposomes were visualized under a transmission electron microscope (JEOL 2100F). Zetasizer Nano ZS (ZEN 3600; Malvern Instruments, Worcestershire, UK) was used to measure NP size and zeta potential. Detailed NP characterization was performed as previously described by Rajendran et al. [55]. Briefly, NPs were diluted 1000-fold in water in order to make homogenous suspension (scatter angle = 90 °C, temperature = 25 °C) and analyzed by placing in a zeta cell (DTS-1060C).

4.4. Drug Concentration Estimation and Encapsulation Efficiency

Drug encapsulated liposomes were lysed by vortexing in 70% ethanol and incubated at 60 °C. Drugs (INH and RIF) were quantified by spectrophotometric absorbance method against a free drug standard curve at 263 nm and 479 nm, respectively [56].

$$\text{Encapsulation efficiency} = (\text{Drug concentration in eluent/drug concentration loaded}) \times 100$$

4.5. Cellular Uptake by Mouse CD271+ BM-MSCs

Briefly, 2 × 10^3 CD271+BM-MSCs and RAW 246.7 cells were seeded in 8-well glass-bottom chambered slide (Ibidi) in D-MEM/F-12 medium with GlutaMAX™-I and DMEM (Thermo Fisher Scientific, Cat. Number 10565018, Cat. Number 12100046, Waltham, MA, USA) serum supplemented media, respectively. After 4–5 h of seeding, prior to imaging, Coumarin6 NP (effective C6 concentration = 2 μg/mL) was added. Live cell imaging was carried out to monitor time lapsed uptake of engineered BTL with Coumarin6 using Nikon Camera Nikon Real-Time Laser Scanning Confocal Microscope, Model A1R. Notedly, C6 is a common fluorescent model drug used for tracking NP uptake [57].

4.6. NP Binding Assay with Bone Chips

Mice were sacrificed to the excise femur bone. Next, bone chips from femur were washed with 1× PBS, incubated with alendronate BTLs encapsulating Coumarin6 (alen BTL C6) or non-targeted liposomes C6 (NT liposomes C6) for 30 min followed by washing. The whole procedure was carried out in 8-well glass-bottom confocal slides. Lastly, each well was added with PBS for visualization through a confocal microscope (Nikon Camera Nikon Real-Time Laser Scanning Confocal Microscope, Model A1R) for the binding of C6 NPs with mouse femur bone chips obtained [35].

4.7. In Vitro Release from BTL-NPs and Tissue Uptake

For in vitro release assay 500 μL of NPs suspension in phosphate-buffered saline (PBS) were added to a dialysis bag (MWCO 12,000–14,000; Sigma-Aldrich, St. Louis, MO, USA). The dialysis bag was placed into 25 mL of PBS (pH 7.4) taken in a dissolution vessel at 37 °C and stirred at 50 rev min^{-1}. At periodic intervals, 500 μL samples were aliquoted from the dialysate and then an equal volume of PBS (pH 7.4) was added. INH and RIF were quantified by measuring absorbance against the free drug standard curve at 263 and 479 nm, respectively. The modified procedure described previously [58] was performed for tissue uptake assay. Briefly, after 8 h of administration of C6 BTL, C6 NT or C6 PBS intravenously, femur and tibia pair were excised followed by chloroform: methanol extraction. The same effective concentration of C6 (150 mg/kg) was administered to all three mice groups. Fluorescence of resulting suspension was determined using Cary Eclipse Fluorescence Spectrophotometer (Agilent) in a quartz cuvette at the emission/excitation of 460/525 nm. The concentration of C6 in the suspension was determined by using the standard curve of C6. % injected a dose of C6 = Conc. of C6/Total injected C6) × 100 per g of organ. Data were represented as % dose injected per organ (per organ in biodistribution study = a pair of tibia and femur from single mice).

4.8. In Vivo Administration and Assessment of BTL-NPs Efficacy

Pathogen-free, 6 weeks female C57bl/6J mice were obtained from the National Institute of Nutrition (Hyderabad, India). Animals were housed and fed according to the standard norms of Animal biosafety level 3 (ABSL3) at Jawaharlal Nehru University, New Delhi. All studies were carried out post-approval from Institution of Animal Ethics Committee (IAEC) and the Institution of Biosafety Committee (IBSC) at Jawaharlal Nehru University, New Delhi.

Mice infections were performed as previously described [13]. Briefly, C57bl/6J female age 6 weeks were infected with H37Rv 2 × 10^6 CFU through tail vein and followed by anti-TB drug therapy (RIF = 0.1 g/L, INH = 0.25 g/L) for 90 days to achieve a status where only Mtb survives in CD271+BM-MSCs while other organs are rendered sterile [11,13]. This model known as the Cornell model has been known for decades where animals treated with anti-Tb drugs wipe out Mtb from lung, liver and spine rendering them sterile, however for no clear reason the disease is relapsed [7,59,60]. It has been shown in mice that this relapse could be due to Mtb present in a novel host cell namely CD271+BM-MSCs [13]. It is postulated that Mtb mesenchymal stem cells may travel from BM to the lung via blood circulation and may cause a relapse of pulmonary TB [11,13]. We have used this Cornell model for targeting Mtb in BM-MSCs. Animals were treated with various treatments including free drugs or BTL-NPs intraperitoneally. The concentration of INH and RIF remained under the safe limits which are 10 mg/kg (standard) for both the drugs [61]. Briefly, INH and RIF average concentrations fell in the range of 2–2.3 mg/mL and 2.66–3.1 mg/mL for their respective NP suspensions (BTL or NT). All liposomes and free drug solutions were diluted to the effective dosage of 3 mg/kg (RIF) or 4 mg/kg (INH) per mice in 150 µL. To be noted, NPs were freshly prepared for each administration and concentrations of drug in each formulation was determined using a standard curve for each drug as mentioned in method Section 4.4. At the end of the regimen, mice were sacrificed and Mtb CFU from CD271+ BM-MSCs sorted cells were enumerated by plating on 7H11 agar plates supplemented with 10% Middlebrook OADC.

4.9. Enumeration of Mtb CFU from CD271+BM-MSCs

The assay was performed as previously described [11–13]. BM cells were aseptically obtained from each group followed by magnetic sorting CD271+ BM-MSCs as described previously [13]. Briefly, an average 4×10^7 bone marrow cells were obtained from a femur pair per mice after RBC lysis. BM cells were then used for performing CD271+ BM-MSCs selection by magnetic sorting (Cat. Number S10467, Life Technologies, Carlsbad, CA, USA; Cat. Number 18554, Stem Cell Technologies, Vancouver, BC, Canada). Sorted cells were lysed followed by plating on 7H11 plates supplemented with 10% Middlebrook OADC.

4.10. TB Relapse Assay

A similar experiment as described in the assessment of BTL-NPs Efficacy (Section 4.8) was carried out. NP treatment regimen was followed by steroid administration (glucocorticoid dexamethasone, intraperitoneal injections of 200 µL of dexamethasone at 10 mg/kg of body weight every 2 days for 4 weeks) and were sacrificed after 4 weeks to look for signs of reactivation. Signs for relapse were (a) granuloma formation in lungs by gross pathology and (b) culture-negative or culture-positive for lungs. Relapse percentage was calculated as follows:

$$\text{Relapse \%} = (\text{No. of mice with culture-positive relapse/total No. of mice}) \times 100$$

4.11. Statistical Analysis

Experimental data were analyzed by GraphPad Prism. Student's *t*-test was performed to compare two groups. Data are expressed as means = +/− SEM; ** $p < 0.001$.

Supplementary Materials: The following are available online at http://www.mdpi.com/2076-0817/9/5/372/s1, Figure S1: Schematic for synthesis of Alendronate conjugation of DSPE-PEG(2000) from DSPE-PEG(2000) Carboxylic Acid and alendronate sodium trihydrate.

Ethical Statement: All experiments related to virulent strain of *Mycobacterium tuberculosis* were carried out inside the BSL3 facility. All animal experiments were conducted in line with "The Indian Animal Ethical Committee" regulations.

Author Contributions: All the authors have contributed according to their expertise in this study. J.G.: Conceptualization, methodology, validation, formal analysis, investigation, visualization, project administration, Original Draft, writing—reviewing and editing. S.M.: Validation, data curation, writing—reviewing and editing. V.R.: Methodology, writing—reviewing and editing. R.B.: Supervision, project administration, funding acquisition. All authors have read and agreed to the published version of the manuscript.

Funding: We extend our sincere thanks to the Department of Biotechnology, Government of India for financial support, specifically for the maintenance and upgradation of BSL-3 and ABSL-3.

Acknowledgments: We are grateful to DBT for providing financial assistance for the BSl-3 work. We would like to thank Gajender Saini (TEM) and A. K. Sahu (Confocal Microscope), Advanced Instrumental Research Facility (AIRF), JNU We also would like to thank Shashank Taxak (confocal image analysis).

Conflicts of Interest: The authors report no conflicts of interest in this work.

References

1. A.O. Global Tuberculosis Report 2019. Available online: https://www.who.int/tb/publications/global_report/en/ (accessed on 20 March 2020).
2. Chaw, L.; Chien, L.C.; Wong, J.; Takahashi, K.; Koh, D.; Lin, R.T. Global trends and gaps in research related to latent tuberculosis infection. *BMC Public Health* **2020**, *20*, 352. [CrossRef] [PubMed]
3. Escalante, P.; Arias-Guillen, M.; Palacios Gutierrez, J.J. New Research Strategies in Latent Tuberculosis Infection. *Arch. Bronconeumol.* **2020**. [CrossRef]
4. Knight, G.M.; McQuaid, C.F.; Dodd, P.J.; Houben, R. Global burden of latent multidrug-resistant tuberculosis: Trends and estimates based on mathematical modelling. *Lancet Infect. Dis* **2019**, *19*, 903–912. [CrossRef]
5. Barry, C.E., 3rd; Boshoff, H.I.; Dartois, V.; Dick, T.; Ehrt, S.; Flynn, J.; Schnappinger, D.; Wilkinson, R.J.; Young, D. The spectrum of latent tuberculosis: Rethinking the biology and intervention strategies. *Nat. Rev. Microbiol.* **2009**, *7*, 845–855. [CrossRef] [PubMed]
6. Turetz, M.L.; Ma, K.C. Diagnosis and management of latent tuberculosis. *Curr. Opin. Infect. Dis.* **2016**, *29*, 205–211. [CrossRef] [PubMed]
7. Scanga, C.A.; Mohan, V.P.; Joseph, H.; Yu, K.; Chan, J.; Flynn, J.L. Reactivation of latent tuberculosis: Variations on the Cornell murine model. *Infect. Immun.* **1999**, *67*, 4531–4538. [CrossRef]
8. Nematollahi, M.H.; Vatankhah, R.; Sharifi, M. Nonlinear adaptive control of tuberculosis with consideration of the risk of endogenous reactivation and exogenous reinfection. *J. Theor. Biol.* **2020**, *486*, 110081. [CrossRef]
9. Dippenaar, A.; De Vos, M.; Marx, F.M.; Adroub, S.A.; van Helden, P.D.; Pain, A.; Sampson, S.L.; Warren, R.M. Whole genome sequencing provides additional insights into recurrent tuberculosis classified as endogenous reactivation by IS6110 DNA fingerprinting. *Infect. Genet. Evol.* **2019**, *75*, 103948. [CrossRef]
10. Pai, M. Spectrum of latent tuberculosis—Existing tests cannot resolve the underlying phenotypes. *Nat. Rev. Microbiol.* **2010**, *8*, 242. [CrossRef]
11. Beamer, G.; Major, S.; Das, B.; Campos-Neto, A. Bone marrow mesenchymal stem cells provide an antibiotic-protective niche for persistent viable Mycobacterium tuberculosis that survive antibiotic treatment. *Am. J. Pathol.* **2014**, *184*, 3170–3175. [CrossRef]
12. Das, B.; Kashino, S.S.; Pulu, I.; Kalita, D.; Swami, V.; Yeger, H.; Felsher, D.W.; Campos-Neto, A. CD271(+) bone marrow mesenchymal stem cells may provide a niche for dormant Mycobacterium tuberculosis. *Sci. Transl. Med.* **2013**, *5*, 170ra13. [CrossRef]
13. Garhyan, J.; Bhuyan, S.; Pulu, I.; Kalita, D.; Das, B.; Bhatnagar, R. Preclinical and Clinical Evidence of Mycobacterium tuberculosis Persistence in the Hypoxic Niche of Bone Marrow Mesenchymal Stem Cells after Therapy. *Am. J. Pathol.* **2015**, *185*, 1924–1934. [CrossRef]
14. Fatima, S.; Kamble, S.S.; Dwivedi, V.P.; Bhattacharya, D.; Kumar, S.; Ranganathan, A.; Van Kaer, L.; Mohanty, S.; Das, G. Mycobacterium tuberculosis programs mesenchymal stem cells to establish dormancy and persistence. *J. Clin. Investig.* **2020**, *130*, 655–661. [CrossRef]
15. Shi, Y.; Hu, G.; Su, J.; Li, W.; Chen, Q.; Shou, P.; Xu, C.; Chen, X.; Huang, Y.; Zhu, Z.; et al. Mesenchymal stem cells: A new strategy for immunosuppression and tissue repair. *Cell Res.* **2010**, *20*, 510–518. [CrossRef]
16. Cohen, K.A.; Abeel, T.; Manson McGuire, A.; Desjardins, C.A.; Munsamy, V.; Shea, T.P.; Walker, B.J.; Bantubani, N.; Almeida, D.V.; Alvarado, L.; et al. Evolution of Extensively Drug-Resistant Tuberculosis over Four Decades: Whole Genome Sequencing and Dating Analysis of *Mycobacterium tuberculosis* Isolates from KwaZulu-Natal. *PLoS Med.* **2015**, *12*, e1001880. [CrossRef]
17. Velayati, A.A.; Masjedi, M.R.; Farnia, P.; Tabarsi, P.; Ghanavi, J.; ZiaZarifi, A.H.; Hoffner, S.E. Emergence of new forms of totally drug-resistant tuberculosis bacilli: Super extensively drug-resistant tuberculosis or totally drug-resistant strains in iran. *Chest* **2009**, *136*, 420–425. [CrossRef]
18. Jacobs, R.E.; Gu, P.; Chachoua, A. Reactivation of pulmonary tuberculosis during cancer treatment. *Int. J. Mycobacteriol.* **2015**, *4*, 337–340. [CrossRef]

19. Motta, I.; Calcagno, A.; Bonora, S. Pharmacokinetics and pharmacogenetics of anti-tubercular drugs: A tool for treatment optimization? *Expert Opin. Drug Metab. Toxicol.* **2018**, *14*, 59–82. [CrossRef]
20. Singh, D.; Goel, D.; Bhatnagar, R. Recombinant L7/L12 protein entrapping PLGA (poly lactide-co-glycolide) micro particles protect BALB/c mice against the virulent *B. abortus* 544 infection. *Vaccine* **2015**, *33*, 2786–2792. [CrossRef]
21. Goel, D.; Rajendran, V.; Ghosh, P.C.; Bhatnagar, R. Cell mediated immune response after challenge in Omp25 liposome immunized mice contributes to protection against virulent Brucella abortus 544. *Vaccine* **2013**, *31*, 1231–1237. [CrossRef]
22. Mata-Espinosa, D.; Molina-Salinas, G.M.; Barrios-Payan, J.; Navarrete-Vazquez, G.; Marquina, B.; Ramos-Espinosa, O.; Bini, E.I.; Baeza, I.; Hernandez-Pando, R. Therapeutic efficacy of liposomes containing 4-(5-pentadecyl-1,3,4-oxadiazol-2-yl)pyridine in a murine model of progressive pulmonary tuberculosis. *Pulm. Pharmacol. Ther.* **2015**, *32*, 7–14. [CrossRef]
23. Manish, M.; Rahi, A.; Kaur, M.; Bhatnagar, R.; Singh, S. A single-dose PLGA encapsulated protective antigen domain 4 nanoformulation protects mice against *Bacillus anthracis* spore challenge. *PLoS ONE* **2013**, *8*, e61885. [CrossRef]
24. Malik, A.; Gupta, M.; Mani, R.; Gogoi, H.; Bhatnagar, R. Trimethyl Chitosan Nanoparticles Encapsulated Protective Antigen Protects the Mice against Anthrax. *Front. Immunol.* **2018**, *9*, 562. [CrossRef]
25. Gogoi, H.; Mani, R.; Bhatnagar, R. A niosome formulation modulates the Th1/Th2 bias immune response in mice and also provides protection against anthrax spore challenge. *Int. J. Nanomed.* **2018**, *13*, 7427–7440. [CrossRef]
26. Jahagirdar, P.S.; Gupta, P.K.; Kulkarni, S.P.; Devarajan, P.V. Intramacrophage delivery of dual drug loaded nanoparticles for effective clearance of *Mycobacterium tuberculosis*. *J. Pharm. Sci.* **2020**. [CrossRef]
27. Zhu, X.; Radovic-Moreno, A.F.; Wu, J.; Langer, R.; Shi, J. Nanomedicine in the Management of Microbial Infection—Overview and Perspectives. *Nano Today* **2014**, *9*, 478–498. [CrossRef]
28. Nisini, R.; Poerio, N.; Mariotti, S.; De Santis, F.; Fraziano, M. The Multirole of Liposomes in Therapy and Prevention of Infectious Diseases. *Front. Immunol.* **2018**, *9*, 155. [CrossRef]
29. Abu Lila, A.S.; Ishida, T. Liposomal Delivery Systems: Design Optimization and Current Applications. *Biol. Pharm. Bull.* **2017**, *40*, 1–10. [CrossRef]
30. Allen, T.M.; Cullis, P.R. Liposomal drug delivery systems: From concept to clinical applications. *Adv. Drug Deliv. Rev.* **2013**, *65*, 36–48. [CrossRef]
31. Sawant, R.R.; Torchilin, V.P. Challenges in development of targeted liposomal therapeutics. *AAPS J.* **2012**, *14*, 303–315. [CrossRef]
32. Poste, G.; Bucana, C.; Raz, A.; Bugelski, P.; Kirsh, R.; Fidler, I.J. Analysis of the fate of systemically administered liposomes and implications for their use in drug delivery. *Cancer Res.* **1982**, *42*, 1412–1422.
33. Sou, K.; Goins, B.; Takeoka, S.; Tsuchida, E.; Phillips, W.T. Selective uptake of surface-modified phospholipid vesicles by bone marrow macrophages in vivo. *Biomaterials* **2007**, *28*, 2655–2666. [CrossRef]
34. Sou, K.; Goins, B.; Leland, M.M.; Tsuchida, E.; Phillips, W.T. Bone marrow-targeted liposomal carriers: A feasibility study in nonhuman primates. *Nanomedicine* **2010**, *5*, 41–49. [CrossRef]
35. Swami, A.; Reagan, M.R.; Basto, P.; Mishima, Y.; Kamaly, N.; Glavey, S.; Zhang, S.; Moschetta, M.; Seevaratnam, D.; Zhang, Y.; et al. Engineered nanomedicine for myeloma and bone microenvironment targeting. *Proc. Natl. Acad. Sci. USA* **2014**, *111*, 10287–10292. [CrossRef]
36. Bhardwaj, A.; Grobler, A.; Rath, G.; Goyal, A.K.; Jain, A.K.; Mehta, A. Pulmonary Delivery of Anti-Tubercular Drugs Using Ligand Anchored pH Sensitive Liposomes for the Treatment of Pulmonary Tuberculosis. *Curr. Drug Deliv.* **2016**, *13*, 909–922. [CrossRef]
37. Franklin, R.K.; Marcus, S.A.; Talaat, A.M.; KuKanich, B.K.; Sullivan, R.; Krugner-Higby, L.A.; Heath, T.D. A Novel Loading Method for Doxycycline Liposomes for Intracellular Drug Delivery: Characterization of in Vitro and in Vivo Release Kinetics and Efficacy in a J774A.1 Cell Line Model of *Mycobacterium smegmatis* Infection. *Drug Metab. Dispos.* **2015**, *43*, 1236–1245. [CrossRef]
38. Singh, J.; Garg, T.; Rath, G.; Goyal, A.K. Advances in nanotechnology-based carrier systems for targeted delivery of bioactive drug molecules with special emphasis on immunotherapy in drug resistant tuberculosis—A critical review. *Drug Deliv.* **2016**, *23*, 1676–1698. [CrossRef]

39. Labana, S.; Pandey, R.; Sharma, S.; Khuller, G.K. Chemotherapeutic activity against murine tuberculosis of once weekly administered drugs (isoniazid and rifampicin) encapsulated in liposomes. *Int. J. Antimicrob. Agents* **2002**, *20*, 301–304. [CrossRef]
40. Deol, P.; Khuller, G.K.; Joshi, K. Therapeutic efficacies of isoniazid and rifampin encapsulated in lung-specific stealth liposomes against Mycobacterium tuberculosis infection induced in mice. *Antimicrob. Agents Chemother.* **1997**, *41*, 1211–1214. [CrossRef]
41. Pandey, R.; Sharma, S.; Khuller, G.K. Lung specific stealth liposomes as antitubercular drug carriers in guinea pigs. *Indian J. Exp. Biol.* **2004**, *42*, 562–566.
42. Farrell, K.B.; Karpeisky, A.; Thamm, D.H.; Zinnen, S. Bisphosphonate conjugation for bone specific drug targeting. *Bone Rep.* **2018**, *9*, 47–60. [CrossRef]
43. Adjei, I.M.; Temples, M.N.; Brown, S.B.; Sharma, B. Targeted Nanomedicine to Treat Bone Metastasis. *Pharmaceutics* **2018**, *10*, 205. [CrossRef]
44. Adachi, J.D. Alendronate for osteoporosis. Safe and efficacious nonhormonal therapy. *Can. Fam. Physician* **1998**, *44*, 327–332.
45. Chou, M.Y.; Yan, D.; Jafarov, T.; Everett, E.T. Modulation of murine bone marrow-derived CFU-F and CFU-OB by in vivo bisphosphonate and fluoride treatments. *Orthod. Craniofac. Res.* **2009**, *12*, 141–147. [CrossRef]
46. La-Beck, N.M.; Liu, X.; Shmeeda, H.; Shudde, C.; Gabizon, A.A. Repurposing amino-bisphosphonates by liposome formulation for a new role in cancer treatment. *Semin. Cancer Biol.* **2019**. [CrossRef]
47. Devogelaer, J.P. A risk-benefit assessment of alendronate in the treatment of involutional osteoporosis. *Drug Saf.* **1998**, *19*, 141–154. [CrossRef]
48. Scheen, A.J. Drug clinics. The drug of the month. Alendronate (Fosamax). *Rev. Med. Liege* **1998**, *53*, 220–222.
49. Sou, K.; Goins, B.; Oyajobi, B.O.; Travi, B.L.; Phillips, W.T. Bone marrow-targeted liposomal carriers. *Expert Opin. Drug Deliv.* **2011**, *8*, 317–328. [CrossRef]
50. Marques-Gallego, P.; de Kroon, A.I. Ligation strategies for targeting liposomal nanocarriers. *Biomed. Res. Int.* **2014**, *2014*, 129458. [CrossRef]
51. Yu, B.; Tai, H.C.; Xue, W.; Lee, L.J.; Lee, R.J. Receptor-targeted nanocarriers for therapeutic delivery to cancer. *Mol. Membr. Biol.* **2010**, *27*, 286–298. [CrossRef]
52. Pinheiro, M.; Lucio, M.; Lima, J.L.; Reis, S. Liposomes as drug delivery systems for the treatment of TB. *Nanomedicine* **2011**, *6*, 1413–1428. [CrossRef]
53. Tandel, N.; Joseph, A.Z.; Joshi, A.; Shrama, P.; Mishra, R.P.; Tyagi, R.K.; Bisen, P.S. An evaluation of liposome-based diagnostics of pulmonary and extrapulmonary tuberculosis. *Expert Rev. Mol. Diagn* **2020**. [CrossRef]
54. Gupta, R.; Rajendran, V.; Ghosh, P.C.; Srivastava, S. Assessment of anti-plasmodial activity of non-hemolytic, non-immunogenic, non-toxic antimicrobial peptides (AMPs LR14) produced by *Lactobacillus plantarum* LR/14. *Drugs R. D* **2014**, *14*, 95–103. [CrossRef]
55. Rajendran, V.; Rohra, S.; Raza, M.; Hasan, G.M.; Dutt, S.; Ghosh, P.C. Stearylamine Liposomal Delivery of Monensin in Combination with Free Artemisinin Eliminates Blood Stages of *Plasmodium falciparum* in Culture and *P. berghei* Infection in Murine Malaria. *Antimicrob. Agents Chemother.* **2015**, *60*, 1304–1318. [CrossRef] [PubMed]
56. Asadpour-Zeynali, K.; Saeb, E. Simultaneous Spectrophotometric Determination of Rifampicin, Isoniazid and Pyrazinamide in a Single Step. *Iran. J. Pharm. Res.* **2016**, *15*, 713–723.
57. Rivolta, I.; Panariti, A.; Lettiero, B.; Sesana, S.; Gasco, P.; Gasco, M.R.; Masserini, M.; Miserocchi, G. Cellular uptake of coumarin-6 as a model drug loaded in solid lipid nanoparticles. *J. Physiol. Pharmacol.* **2011**, *62*, 45–53.
58. Ramanlal Chaudhari, K.; Kumar, A.; Megraj Khandelwal, V.K.; Ukawala, M.; Manjappa, A.S.; Mishra, A.K.; Monkkonen, J.; Ramachandra Murthy, R.S. Bone metastasis targeting: A novel approach to reach bone using Zoledronate anchored PLGA nanoparticle as carrier system loaded with Docetaxel. *J. Control. Release* **2012**, *158*, 470–478. [CrossRef]
59. Radaeva, T.V.; Kondratieva, E.V.; Sosunov, V.V.; Majorov, K.B.; Apt, A. A human-like TB in genetically susceptible mice followed by the true dormancy in a Cornell-like model. *Tuberculosis* **2008**, *88*, 576–585. [CrossRef]

60. de Wit, D.; Wootton, M.; Dhillon, J.; Mitchison, D.A. The bacterial DNA content of mouse organs in the Cornell model of dormant tuberculosis. *Tuber Lung Dis.* **1995**, *76*, 555–562. [CrossRef]
61. Gupta, S.; Tyagi, S.; Almeida, D.V.; Maiga, M.C.; Ammerman, N.C.; Bishai, W.R. Acceleration of tuberculosis treatment by adjunctive therapy with verapamil as an efflux inhibitor. *Am. J. Respir. Crit. Care Med.* **2013**, *188*, 600–607. [CrossRef]

Review

Intelligent Mechanisms of Macrophage Apoptosis Subversion by *Mycobacterium*

Abualgasim Elgaili Abdalla [1,2,*], Hasan Ejaz [1], Mahjoob Osman Mahjoob [2], Ayman Ali Mohammed Alameen [1,3], Khalid Omer Abdalla Abosalif [1,2], Mohammed Yagoub Mohammed Elamir [1,2] and Mohammed Alsadig Mousa [2]

1 Department of Clinical Laboratory Sciences, College of Applied Medical Sciences, Jouf University, Al Jouf 2014, Saudi Arabia; hetariq@ju.edu.sa (H.E.); aaalameen@ju.edu.sa (A.A.M.A.); koabosalif@ju.edu.sa (K.O.A.A.); myelamir@ju.edu.sa (M.Y.M.E.)

2 Department of Medical Microbiology, Faculty of Medical Laboratory Sciences, Omdurman Islamic University, Omdurman 14415, Sudan

3 Department of Chemical Pathology, Faculty of Medical Laboratory Sciences, University of Khartoum, Khartoum 11081, Sudan

* Correspondence: aealseddig@ju.edu.sa; Tel.: +966-539112018

Received: 25 February 2020; Accepted: 15 March 2020; Published: 16 March 2020

Abstract: Macrophages are one of the first innate defense barriers and play an indispensable role in communication between innate and adaptive immune responses, leading to restricted *Mycobacterium tuberculosis* (*Mtb*) infection. The macrophages can undergo programmed cell death (apoptosis), which is a crucial step to limit the intracellular growth of bacilli by liberating them into extracellular milieu in the form of apoptotic bodies. These bodies can be taken up by the macrophages for the further degradation of bacilli or by the dendritic cells, thereby leading to the activation of T lymphocytes. However, *Mtb* has the ability to interplay with complex signaling networks to subvert macrophage apoptosis. Here, we describe the intelligent strategies of *Mtb* inhibition of macrophages apoptosis. This review provides a platform for the future study of unrevealed *Mtb* anti-apoptotic mechanisms and the design of therapeutic interventions.

Keywords: Mycobacterium; macrophage; apoptosis; effector; cytokine; microRNA

1. Introduction

Tuberculosis (TB) is a chronic infectious disease caused by *Mycobacterium tuberculosis* (*Mtb*), which affects 10 million people globally. It was one of the top 10 infectious diseases in 2017, with an estimated 1.6 million deaths worldwide, which dropped to 1.5 million in 2018 [1]. *Mtb* is considered a sinful intercellular pathogen that is capable of surviving and replicating within the hostile microenvironment of macrophages and other cellular niches [2]. Macrophages are at the frontline of the innate defense, encounter the pathogens, and play a critical role in the containment of bacilli infection through triggering inflammation, digestion of bacilli, and inducing adaptive immune responses [3]. However, *Mtb* can subvert macrophage responses such as apoptosis in order to establish a persistent lifestyle [4]. Thus, understanding the underlying mechanisms behind *Mtb* manipulation of macrophages apoptosis is decisive for treatment interventions and TB vaccines.

Apoptosis is a highly coordinated and regulated process of the programmed cell death form in which dying cell components and *Mtb* are enclosed within cytoplasmic membranes and liberated from the cell called apoptotic bodies [5]. The apoptosis of macrophages is a crucial part of the innate host defense against *Mtb* through restricting the intracellular growth of bacilli. The *Mtb* infected macrophage undergoes apoptosis in order to release the *Mtb* into the extracellular milieu, thereby leading to the activation of dendritic cells and triggering of robust adaptive immune responses [6,7].

Nevertheless, *Mtb* has evolved multiple mechanisms to revoke macrophages apoptosis, which are discussed in the current review.

2. Anti-Apoptotic Determinants of Mycobacterium

Mtb has a wide variety of effector molecules that actively block macrophage apoptotic pathways. In this context, cell wall-associated glycolipids, such as lipoarabinomannan (LAM) and mannosylated LAM (ManLAM), are important virulence determinants that have been found to modulate macrophages apoptosis [8,9]. ManLAM can thwart *Mtb* mediating B10R macrophage apoptosis by inhibiting calcium influx and its intracellular signaling. Ablation of calcium-signaling leads to the inhibition of caspase-1 activity, alteration of mitochondrial membrane permeability, as well as induced upregulation of anti-apoptotic, namely, B-cell lymphoma 2 (Bcl2) (Table 1) [8]. LAM can also block human leukemia monocytic cell (THP-1) apoptosis by activating phosphatidylinositol 3-kinase (PI3K), which in turn activates serine/threonine kinase (Akt) that suppresses pro-apoptotic factor Bad (Table 1) [9].

Mtb tyrosine phosphatase (PtpA) is a secreted protein that plays a decisive role in the pathogenesis and survival of *Mtb* within macrophages via interplay with multiple host signaling pathways [10]. PtpA knockout *Mtb* strain induces a high level of caspase-3 expression and promotes its activity in THP-1 macrophages when compared with wild *Mtb* strains. PtpA can block caspase-3 activity by dephosphorylation of human glycogen synthase kinase 3 (GSK3) (Table 1) [11]. Similarly, PtpA can attenuate differentiated U937 macrophages apoptosis in response to Bacillus Calmette–Guérin (BCG) strain infections. PtpA can abolish the ubiquitin ligase of the tripartite motif-containing (TRIM) protein activity, which is required for pro-caspase-3 cleavage [12]. Noticeable, another *Mtb* secreted tyrosine phosphatase (MptpB) is a decisive virulence factor that can modulate macrophage responses and enhance *Mtb* intracellular survival. RAW264.7 cells expressing MptpB displayed much less apoptosis than the control cells when infected with BCG strains. Furthermore, interferon-gamma (IFN-γ)-stimulated RAW264.7 cells expressing MptpB show a high viability rate and low levels of cleaved caspase-3 in response to BCG infection when compared RAW264.7 cells-harboring empty vector [13]. Taken together, the anti-apoptotic mechanisms of *Mtb* tyrosine phosphatases are dependent on the inhibition of caspase-3 activation (Table 1).

Mtb has 11 types of serine/threonine protein kinases (PknA through PknL) that play a significant role with respect to bacterial adaptation in hostile environments, both in vivo and in vitro [14]. Among them, pknE has anti-apoptotic activity and is indispensable for *Mtb* survival inside the macrophage. PknE deleted *Mtb* strains induce higher levels of THP-1 macrophage apoptosis than the wild type strains and can suppress oxidative stress-inducing cellular apoptosis [15,16]. PknE can modulate the expression of multiple apoptotic molecules and reduce the expression of pro-apoptotic factors, including P53, Bax, and TNF-α, while increasing the expression of anti-apoptotic factor Mcl-1 (Table 1). Furthermore, pknE can promote the phosphorylation of Akt [16], which in turn suppresses the pro-apoptotic factor, i.e., Bad protein [9].

Mtb NADH-ubiquinone oxidoreductase subunit G (nuoG), a subunit of type-1 NADH dehydrogenase (NADH-1), was demonstrated to suppressed THP-1 and murine bone marrow-derived macrophages (BMDMs) apoptosis and promoted the pathogenesis of bacilli in a mice infection model [17]. NuoG knockout *Mtb* strains were less virulent and lost their ability to abrogate macrophage apoptosis in comparison with control strains. The ability of nuoG deleted *Mtb* strains to provoke macrophage apoptosis was significantly reversed upon treatment with caspase-3 and caspase-8 inhibitors, or through the ablation of TNF-α signaling. NuoG can suppress NADPH oxidase (NOX2)-inducing of ROS production leading to inhibition of TNF-α secretion (Table 1) [18]. A recent study demonstrates that the deletion of *Mtb* TNF-α-suppressing genes results in increased macrophage apoptosis and promotes a cell-mediated immune response [19]. Therefore, targeting nuoG might improve the immune responses and control the intracellular growth of *Mtb*.

Mtb nucleoside diphosphate kinase (Ndk) acts as a small GTPase inhibitor that can deactivate GTPase Rac1 and, subsequently, inhibit NOX2 assembly and ROS production. Ndk has been demonstrated to block murine macrophage apoptosis by abolishing ROS-mediated caspase-3 cleavage (Table 1) [20].

Mtb isocitrate lyase (Icl) is crucial for bacilli viability during latent infection [21]. The recombinant *M. smegmatis*-expressing IcI enhances survival within RAW264.7 murine macrophages and inhibits apoptosis in comparison with *M. smegmatis*-harboring empty vector [22]. Nevertheless, the underlying mechanisms by which IcI subverts macrophage apoptosis are yet be determined.

Mtb PE_PGRS family proteins are the most abundant cell wall anchored proteins that are implicated in bacilli adhesion, invasion, and survival within macrophages and dendritic cells [23]. PE_PGRS62, PE_PGRS41, and PE_PGRS18 contribute actively to revoke macrophage apoptosis [24–26].

The ectopic expression of PE_PGRS62 in *M. smegmatis* enhances its survival within THP-1 macrophages and reduces cell apoptosis. It has been observed that PE_PGRS62 can inhibit the endoplasmic reticulum (ER) stress response (Table 1) by downregulating C/EBP homologous protein (CHOP) and the 78-kDa glucose-regulated protein (GRP78/Bip), which are essential pro-apoptotic factors in response to stress conditions [25]. Likewise, *M. smegmatis*-expressing PE_PGRS41 induces much lower THP-1 macrophage apoptosis than the *M. smegmatis* expressing vector. PE_PGRS41 can decrease the cleavage levels of caspase-9 and caspase-3 (Table 1) [24]. PE_PGRS18 also inhibits THP-1 apoptosis. However the underlying molecular mechanisms are still unknown [26].

Mtb secreted proteins encoded by *Rv3654c* and *Rv3655c* genes express within the cytoplasm of *Mtb*-infected macrophages, suggesting that they may have a role in *Mtb* interplays with host cell signaling [27]. These proteins abrogate the extrinsic pathway-mediated U937 cells apoptosis during *Mtb* infection. They interact and cleave the polypyrimidine tract binding Protein-associated Splicing Factor (PSF) result in the deactivation of caspase-8 (Table 1) [27]. Similarly, Rv3033 is a secreted protein that is crucial for *Mtb* viability within macrophages [28]. Rv3033 has potent anti-apoptotic activity by suppressing the intrinsic apoptotic pathway. *M. smegmatis* expressing Rv3033 inhibits murine BMDM apoptosis compared with control strains. Rv3033 can actively block the translocation of pro-apoptotic Bax protein into mitochondria and cytochrome c into cytoplasm, thereby leading to suppress the caspase-9 activation (Table 1) [29]. Likewise, Rv3365c is another hypothetical protein that is an important *Mtb* anti-apoptotic effector. The deletion of Rv3365c significantly reduces the ability of *Mtb* to suppress U937 cells apoptosis in comparison with wild type strains. Rv3365c can revoke cellular apoptosis by interacting and inhibiting host cell membrane-bound serine protease cathepsin G and its downstream activation of caspase-1(Table 1) [30].

Mtb enhanced intracellular survival (Eis) protein is an important modulator of macrophage apoptosis, autophagy, and inflammatory cytokines production. In the context of apoptosis, Eis knockout *Mtb* strain significantly increases the murine BMDM when compared with wild type or complemented strains. Eis can repress JNK signaling leading to inhibition of ROS production (Table 1) [31].

Mtb stress response proteins play a significant role in bacterial resistance to the harsh milieu in both *in vitro* and in vivo [32,33]. *Mtb* SigH is a stress response regulon that controls the expression of multiple *Mtb* genes in response to oxidative and heat-stressors, as well as to host microenvironments [34,35]. Lower apoptosis of primary rhesus macaque bone marrow-derived macrophage (Rh-BMDM) was observed upon infection with SigH knockdown *Mtb* strains compared with wild *Mtb* strains [33]. SigH regulon can abolish cellular apoptosis through boosting prostaglandin synthetase 2 (PTGS2) expression, which in turn inhibits the P53-dependent apoptotic pathway (Table 1) [33].

Mtb Cpn60.2 (GroEL2) is a member of heat shock proteins found in the secreted form within the cytosol of macrophage [32]. Cpn60.2 can revoke THP-1 macrophage apoptosis through interacting with, and enhancing the stability of human mitochondrial Hsp70, which is (Table 1) an important anti-apoptotic factor [32] also known as mortalin [36].

Mtb acpM is a meromycolate extension acyl carrier protein involved in cell wall mycolic acid biosynthesis and a crucial drug target to eradicate multi-drugs-resistant *Mtb* strains (MDR) [37]. *M. smegmatis* expressing-acpM (Msm-acpM) can enhance the viability of BMDM in comparison with cells infected with *M. smegmatis,* harboring an empty vector. AcpM reduces ROS production by suppressing c-Jun N-terminal Kinase (JNK) signaling (Table 1) [38].

Mtb LpqT is a cell wall anchored lipoprotein that modulates macrophage responses by engaging with Toll-Like Receptor-2 (TLR-2) [39]. LpqT deleted *M. smegmatis* strains significantly induce higher levels of RAW264.7 macrophage apoptosis compared with control strains. LpqT could decrease the cleaved levels of caspase-3 by blocking the TLR-2 signaling pathway (Table 1) [39]. However, the previous study demonstrated that LpqT could increase the apoptosis of THP-1 macrophage and monocyte-derived macrophages (MDM) via agonist TLR-2 signaling [40]. Therefore, the exact role of *Mtb* LpqT in the regulation of macrophage apoptosis in the context of its interaction with TLR-2 remains controversial. It has been demonstrated that lipoprotein MPT83 can induce protective immunity against *Mtb* infection via promoting macrophage apoptosis [41].

EspR is an *Mtb* specific protein and a key regulatory protein of the ESX-1 secretion system and many other genes that are involved in *Mtb* pathogenesis [42]. Enforced expression of EspR in RAW264.7 macrophage leading to suppressed BCG mediating apoptosis. EspR can significantly decrease the cleavage of both caspase-3 and caspase-8 (Table 1) when compared with RAW264.7 harboring empty-vector. The mechanisms underlying caspase inhibition was due to EspR interfering with TLRs signaling by directly interacting with myeloid differentiated protein-88 (MyD88) [43].

Table 1. Anti-apoptotic effectors of *Mycobacterium.*

Effector	Cell Model	Mechanisms	Outcome	References
ManLAM	B10R	Blocks Ca^{+2} influx to the cells.	Inhibition of caspase-1 cleavage, alter mitochondrial membrane permeability and upregulate Bcl-2	Rojas et al., 2000
LAM	THP-1	Activation of PI3K signaling	Suppression of Bad	Maiti et al., 2001
PtpA	THP-1	Dephosphorylation of GSK3	Inhibition of caspase-3 cleavage	Poirier et al., 2014
	U937	Suppress ubiquitin ligase activity of the TRIM protein	Inhibition of caspase-3 cleavage	Wang et al., 2016
MptpA	RAW264.7	Reduction of P53 levels	Inhibition of caspase-3 cleavage	Fan et al., 2018
PknE	THP-1	Phosphorylation of Akt	Inhibition of Bad	Kumar and Narayanan et al., 2012
		Inhibit the expression of pro-apoptotic factors, including P53, TNF-α and Bax	Inhibition of caspase-3 activation	
		Promote the anti-apoptotic factor Mcl-1 expression	Block Bax mitochondrial translocation	
NuoG	THP-1 and BMDM	Blocks of NADPH oxidase mediating ROS production	Inhibition of TNF-α production	Miller et al., 2010
Ndk	RAW264.7	Inhibit NOX2 assembly and ROS production	Inhibition of caspase-3 activation	Sun et al., 2013
Icl	RAW264.7	Unknown	Unknown	Li et al., 2008
PE_PGRS62	THP-1	Suppression of pro-apoptotic stress-response genes expressions such as CHOP and GRP78/Bip	Inhibit endoplasmic reticulum (ER) stress response	Long et al., 2019
PE_PGRS41	THP-1	Uncertain	Reduction the cleavage level of caspase 3 and 9	Deng et al., 2017
PE_PGRS18	THP-1	Unknown	Unknown	Yang et al., 2017

Table 1. *Cont.*

Effector	Cell Model	Mechanisms	Outcome	References
Rv3654c Rv3655c	U937	Degrade the polypyrimidine tract binding PSF	Suppression of caspase-8 activation	Danelishvili et al., 2010
Rv3033	RAW264.7 and murine BMDM	Abolish translocation of Bax into mitochondria and cytochrome c into cytoplasm	Suppression of caspase-9 activation	Zhang et al., 2018
Rv3365c	U937	Inhibit serine cathepsin G	Suppression of caspase-1	Danelishvili et al., 2012
Eis	Murine BMDM	Block the JNK signaling	Inhibition of ROS production	Shin et al., 2010
SigH	Rh-BMDM	Promote prostaglandin synthetase-2 expression	Inhibition P53 dependent pathway	Dutta et al., 2012
AcpM	Murine BMDM	Suppress JNK signaling	Reduction of ROS production	Paik et al., 2019
LpqT	RAW264.7	Antagonized TLR-2 signaling	Inhibition of caspase-3 cleavage	Li et al., 2018
EspR	RAW264.7	Block TLR signaling	Inhibition of caspase-8 and 3 cleavage	Jin et al., 2019

PI3K, phosphatidylinositol 3-kinase; GSK3, glycogen synthetase-3; TRIM, tripartite motif; CHOP, C/EBP homologous protein; GRP78/Bip, 78-kDa glucose-regulated protein; PSF, Protein-associated Splicing Factor; TLR, Toll-Like Receptor; JNK, c-Jun N-terminal Kinase; BMDM, bone marrow-derived macrophages; Rh, Rhesus.

3. Mycobacterium Thwarts Macrophage Apoptosis by Inducing Anti-Apoptotic Cytokines

Cytokines are inducible, soluble immune mediators responsible for orchestrating various and complex immunological processes, including immune cells crosstalk and apoptosis [44]. *Mtb* can differentially induce cytokines secretion to manipulate macrophage apoptosis. An earlier study showed that *Mycobacterium* could inhibit macrophage apoptosis by robust inducing interleukin-10 (IL-10) production, leading to the increased expression of Bcl2 and decreasing pro-apoptotic factors such as cleaved caspase-1, P53, nitric oxide (NO), and TNF-α production (Figure 1) [45]. Transgenic mice expressing human IL-10 in antigen-presenting cells (APCs) infected with *M. avium* displayed lower levels of macrophage apoptosis, which was accompanied by decreased TNF-α and NO production [46]. IL-10 also inhibits *Mtb* inducing alveolar macrophages (AM) apoptosis by abolishing TNF-α production. In this regard, IL-10 can increase the levels of B-cell lymphoma 3 protein (Bcl-3), which in turn antagonizes TNF-α production via deactivating NF-kappaB nuclear signaling [47].

IL-17A, a cytokine secreted in high amount by T helper-17 (Th17) in response to *Mtb* infection and actively contributes to TB pathogenesis [48]. It has been shown that the exogenous addition of IL-17A can significantly enhance BCG strains and *Mtb* viability within murine BMDM when compared with bacterial intracellular growth in the absence of IL-17A. This cytokine can blockade BMDM apoptosis upon BCG infection by suppression of intrinsic apoptotic pathway through interfering with nuclear translocation of P53, thereby leading to the upregulation of Bcl2 expression and downregulation of Bax expression, as well as abolishing cytochrome c released and caspase-3 activation (Figure 1) [49].

Epstein–Barr virus-induced gene 3 (EBI3), is a subunit of anti-inflammatory cytokines, namely IL-27 and IL-35. It was found that EBI3 is produced in a significant amount in CD14+ macrophages isolated from TB patients compared to cells from healthy donors. Furthermore, it has been demonstrated that EBI3 production and accumulation were markedly increased upon the challenge of murine peritoneal macrophages with *Mtb*, suggesting that it may contribute to the immunoevasion of macrophage responses. *Mtb* and BCG strains shown to induce significant apoptosis of murine EBI3 deleted macrophages in comparison with the deaths of wild type cells. EB13 abrogates the extrinsic apoptotic pathway via suppression of the cleavage of caspase-3 and caspase-8 (Figure 1) [50]. Taken together, *Mtb* can stimulate IL-10, IL-17A, and EBI3 secretion to frustrate macrophage apoptosis. Thus, complementary to antibodies blocking the action of these cytokines, anti-tuberculous drugs can hasten the eradication of bacilli and shortage of treatment regimens.

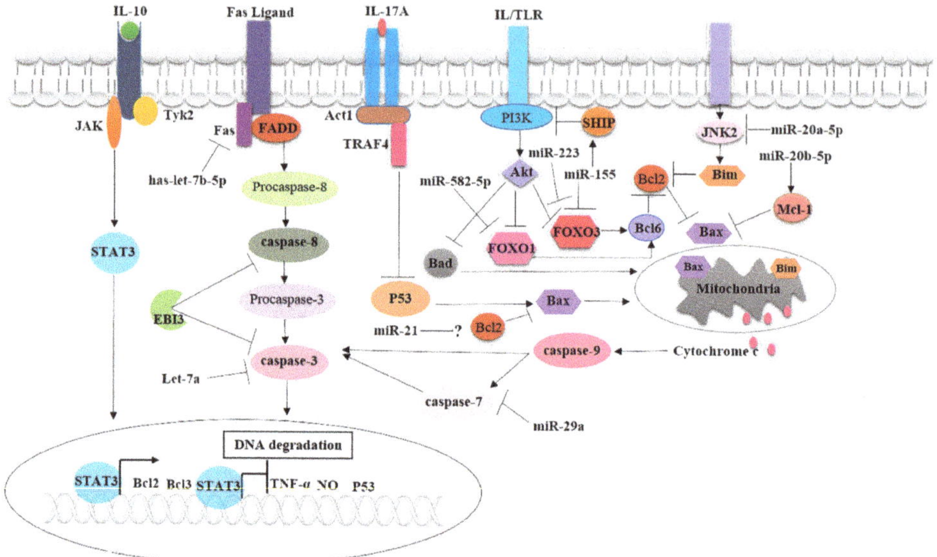

Figure 1. Mycobacterium evades macrophage apoptosis by induction of anti-apoptotic cytokines and miRNAs expression. Three cytokines, namely, IL-10, IL-17A, and EBI3, were demonstrated to play an anti-apoptotic role during *Mtb* infection. IL-10 can mediate the expression of anti-apoptotic Bcl2 and Bcl3 protein and suppresses the production of pro-apoptotic factors, including TNF-α, nitric oxide (NO), and P53. IL-17A signaling leads to suppression of P53 mediated Bax mitochondrial translocation. EBI3 can directly inhibit caspase-8 and caspase-3 activation. Nine anti-apoptotic miRs were found selectively regulated by *Mtb*. Has-let-7b-5p can target Fas leading to inhibition of downstream activation of caspase-3. Let-7a and miR-29a are targeting caspase-3 and caspase-7, respectively. MiR-155 can abolish expression of SH2 domain-containing inositol 5-phosphatase 1(SHIP1), leading to activation of phosphatidylinositol 3-kinase (PI3K) signaling-mediated inhibition of pro-apoptotic factors including Bad, FOXO-1 and 3. In addition, miR-155 can directly targeting and suppressing the expression of FOXO-3. MiR-582-5p can target the mRNA of FOXO-1 and block its expression. MiR-20a-5p can abrogate the activation of pro-apoptotic factor Bim by targeting the JNK-2 signaling pathway. However, miR-20b-5p can enhance the expression of anti-apoptotic factor Mcl-1, which can mediate block Bax mitochondrial translocation. The role of miR-21 in regulating anti-apoptotic Bcl-2 remains controversial.

4. *Mtb* can Suppress Apoptosis by Regulating microRNAs Expression

MicroRNAs (miRs) are small noncoding RNAs that play an indispensable role in posttranscriptional regulation of genes expression that involved in various cellular biochemical pathways, including immune signaling pathways in response to pathogenic infections such as *Mtb* [51].

Mtb can selectively regulate miRs expression to defeat macrophages apoptosis in order to enhance its intracellular viability and growth. Consistently, miR-582-5p has been significantly more upregulated in monocytes obtained from active TB patients than the cells from healthy controls. The transfection of THP-1 monocytic cells with miR-582-5p mimics leads to the blocking of apoptosis when compared cells transfected with the negative control. MiR-582-5p can inhibit monocytes apoptosis by directly targeting forkhead box O1 (FOXO1) mRNA and suppressing its translation (Figure 1) [52].

MiR-155 expression significantly upregulated in the peripheral blood monocytes (PBMCs) obtained from patients with active TB disease than the cells from healthy control [53]. The apoptosis of CD4+

monocytes was significantly decreased in active TB patients compared with healthy control, suggesting that miR-155 may abolish cellular apoptosis. The ectopic expression of miR-155 in THP-1 macrophages leading to abrogate BCG-inducing apoptosis via downregulating FOXO3 expression (Figure 1) [53]. Furthermore, *Mtb* infected miR-155 knockdown murine macrophages exhibited a higher level of apoptosis when compared with wild type macrophages infected with *Mtb* [54]. MiR-155 can target the mRNA of SH2 domain-containing inositol 5-phosphatase 1(SHIP1), which was confirmed by increased protein levels in miR-155 deleted macrophages. Increased levels of SHIP1 can result in reduced phosphorylation of Akt, thereby leading to the enhanced activation of Bad, FOXO-1, 3, and caspase-3 (Figure 1) [54].

MiR-223 is expressed significantly more in peripheral macrophages harvested from active TB patients compared with cells from healthy controls and upregulated during the in vitro *Mtb* infection of human macrophages. The forced expression miR-223 in human macrophages significantly reduced the rate of apoptosis in comparison with macrophages transfected with control miR-223 mimics. Mechanistically, miR-223 blocks macrophage apoptosis by targeting FOXO3 and markedly reducing its protein expression (Figure 1) [55].

Let-7e and miR-29a are highly expressed in human monocyte-derived macrophages in response to *M. avium* infection. They block macrophages apoptosis by repressing the expression of caspase-3 and 7, which are direct targets for let-7e and miR-29a, respectively (Figure 1) [56].

MiR-21 expression was upregulated by NF-κB signaling in RAW264.7 murine macrophage triggered by *Mtb* mpt68 protein. It has been demonstrated that miR-21 could inhibit murine macrophage apoptosis by positively regulating anti-apoptotic Bcl-2 expression (Figure 1) [57]. However, recent studies have demonstrated that miR-21 enhances macrophages apoptosis in response to mycobacterium infection by negatively regulating the expression of Bcl-2 [58,59]. Therefore, the role of miR-21 in regulating macrophage apoptosis and Bcl-2 expression remains ambiguous.

MiR-20a-5p expression was found to be downregulated in CD14+ monocytes from active pulmonary TB patients, while it is pronouncedly upregulated by successful anti-TB treatment. The downregulation of miR-20a-5p expression also observed during the *Mtb* infection of THP-1 human macrophages. Blockade of miR-20a-5p expression can accelerate THP-1 macrophage apoptosis and promote mycobacterium survival. MiR-20a-5p ectopically expressed in THP-1 cells led to the decreased expression of pro-apoptotic gene Bim through targeting and abolishing the expression of JNK2 signaling (Figure 1) [60]. Recently, the downregulation of miR-20b-5p expression has been reported upon the *Mtb* infection of RAW264.7 macrophages. The overexpression of miR-20b-5p can promote bacilli survival and cell apoptosis [61]. In this regard, miR-20b-5p can increase macrophage apoptosis by targeting and suppressing the expression of the anti-apoptotic Mcl-1 gene (Figure 1) [61], which is a member of the Bcl-2 protein family.

Hsa-let-7b-5p was significantly upregulated in *Mtb* infected THP-1 macrophages, leading to the subversion of apoptosis. Mechanistically, hsa-let-7b-5p negatively regulates the expression of Fas by directly targeting its mRNA, which can lead to blocking pathways of caspase-3 activation (Figure 1) [62].

5. Conclusions

Macrophages are the first line of innate defense against *Mtb* infection, and the subversion of macrophage apoptosis is considered a hallmark of *Mtb* pathogenesis. *Mtb* is a mystery pathogen that utilizes multiple strategies to abolish apoptotic signaling pathways in order to establish persistent infection. The impairment of macrophage apoptosis leads to the enhanced intracellular survival of bacilli and compromises the cell-mediated immune response. *Mtb* has a broad spectrum of anti-apoptotic effector molecules that direct targeting cellular pro-apoptotic factors or block the signaling that regulates their expression. Some *Mtb* anti-apoptotic effectors, including PE_PGRS18 and IcI, impaired macrophage apoptosis by unrecognized mechanisms. Moreover, the role of *Mycobacterium* lipoprotein (LpqT) in modulating macrophage apoptosis needed to be clarified. Future studies are needed to identify unrevealed mechanisms of *Mtb* anti-apoptotic effectors and to develop methods able to hamper *Mtb* effectors' capability to abrogate macrophage apoptosis.

Mtb can also manipulate macrophage apoptosis by selectively regulating cytokines and miRs expression. It is also important to understand how *Mtb* selectively regulates anti-apoptotic cytokines and miRs expression. The mechanisms by which cytokines and miRs enhance the viability of macrophages are dependent on the aberrant activity of key pro-apoptotic regulators, including P53, TNF-α, Fas, FOXO, and Bad. They promote the activity of anti-apoptotic regulators such as SHIP, Bcl-2, and Mcl-1. Taken together, these insights into the intelligent subversion mechanisms of macrophage apoptosis by *Mtb* elucidate promising and novel therapeutic targets to eliminate the intracellular survival of bacilli and promote an adaptive immune response.

Author Contributions: A.E.A. conceived the idea and drafted the manuscript. M.O.M., A.A.M.A., K.O.A.A., M.Y.M.E., and M.A.M. reviewed the literature and helped in the drafting and analysis of the manuscript. H.E. contributed in drafting and critical revision of the manuscript. All authors have read and agree to the published version of the manuscript.

Funding: This research received no external funding.

Conflicts of Interest: The authors declare no conflict of interest.

References

1. World Health Organization. *Global Tuberculosis Report 2019*; World Health Organization: Geneva, Switzerland, 2019; Available online: https://www.who.int/tb/publications/global_report/en/ (accessed on 17 October 2019).
2. Marakalala, M.J.; Martinez, F.O.; Plüddemann, A.; Gordon, S. Macrophage heterogeneity in the immunopathogenesis of tuberculosis. *Front. Microbiol.* **2018**, *9*, 1028. [CrossRef] [PubMed]
3. Ferraris, D.; Miggiano, R.; Rossi, F.; Rizzi, M. *Mycobacterium tuberculosis* molecular determinants of infection, survival strategies, and vulnerable targets. *Pathogens* **2018**, *7*, 17. [CrossRef] [PubMed]
4. Hmama, Z.; Peña-Díaz, S.; Joseph, S.; Av-Gay, Y. Immunoevasion and immunosuppression of the macrophage by *Mycobacterium tuberculosis*. *Immunol. Rev.* **2015**, *264*, 220–232. [CrossRef] [PubMed]
5. Divangahi, M.; Behar, S.M.; Remold, H. Dying to live: How the death modality of the infected macrophage affects immunity to tuberculosis. *Adv. Exp. Med. Biol.* **2013**, *783*, 103–120. [CrossRef] [PubMed]
6. Weiss, G.; Schaible, U.E. Macrophage defense mechanisms against intracellular bacteria. *Immunol. Rev.* **2015**, *264*, 182–203. [CrossRef] [PubMed]
7. Kelly, D.M.; ten Bokum, A.M.; O'Leary, S.M.; O'Sullivan, M.P.; Keane, J. Bystander macrophage apoptosis after *Mycobacterium tuberculosis* H37Ra infection. *Infect. Immun.* **2008**, *76*, 351–360. [CrossRef] [PubMed]
8. Rojas, M.; García, L.F.; Nigou, J.; Puzo, G.; Olivier, M. Mannosylated lipoarabinomannan antagonizes *Mycobacterium tuberculosis*-induced macrophage apoptosis by altering Ca+ 2-dependent cell signaling. *J. Infect. Dis.* **2000**, *182*, 240–251. [CrossRef]
9. Maiti, D.; Bhattacharyya, A.; Basu, J. Lipoarabinomannan from *Mycobacterium tuberculosis* promotes macrophage survival by phosphorylating bad through a phosphatidylinositol 3-kinase/Akt pathway. *J. Biol. Chem.* **2001**, *276*, 329–333. [CrossRef]
10. Wong, D.; Chao, J.D.; Av-Gay, Y. *Mycobacterium tuberculosis*-secreted phosphatases: From pathogenesis to targets for TB drug development. *Trends Microbiol.* **2013**, *21*, 100–109. [CrossRef]
11. Poirier, V.; Bach, H.; Av-Gay, Y. *Mycobacterium tuberculosis* promotes anti-apoptotic activity of the macrophage by PtpA protein-dependent dephosphorylation of host GSK3α. *J. Biol. Chem.* **2014**, *289*, 29376–29385. [CrossRef]
12. Wang, J.; Teng, J.L.; Zhao, D. The ubiquitin ligase TRIM27 functions as a host restriction factor antagonized by *Mycobacterium tuberculosis* PtpA during mycobacterial infection. *Sci. Rep.* **2016**, *6*, 34827. [CrossRef] [PubMed]
13. Fan, L.; Wu, X.; Jin, C.; Li, F.; Xiong, S.; Dong, Y. MptpB promotes *Mycobacteria* survival by inhibiting the expression of inflammatory mediators and cell apoptosis in macrophages. *Front. Cell Infect. Microbiol.* **2018**, *8*, 171. [CrossRef] [PubMed]
14. Prisic, S.; Husson, R.N. *Mycobacterium tuberculosis* serine/threonine protein kinases. *Microbiol. Spectr.* **2014**, *2*, 1–42. [CrossRef]

15. Jayakumar, D.; Jacobs, W.R.; Narayanan, S. Protein kinase E of *Mycobacterium tuberculosis* has a role in the nitric oxide stress response and apoptosis in a human macrophage model of infection. *Cell Microbiol.* **2008**, *10*, 365–374. [CrossRef]
16. Kumar, D.; Narayanan, S. pknE, a serine/threonine kinase of *Mycobacterium tuberculosis* modulates multiple apoptotic paradigms. *Infect. Genet. Evol.* **2012**, *12*, 737–747. [CrossRef]
17. Velmurugan, K.; Chen, B.; Miller, J.L. *Mycobacterium tuberculosis* nuoG is a virulence gene that inhibits apoptosis of infected host cells. *PLoS Pathog.* **2007**, *3*, e110. [CrossRef]
18. Miller, J.L.; Velmurugan, K.; Cowan, M.J.; Briken, V. The type I NADH dehydrogenase of *Mycobacterium tuberculosis* counters phagosomal NOX2 activity to inhibit TNF-α-mediated host cell apoptosis. *PLoS Pathog.* **2010**, *6*, e1000864. [CrossRef]
19. Olsen, A.; Chen, Y.; Ji, Q.; Zhu, G.; De Silva, A.D.; Vilchèze, C.; Weisbrod, T.; Li, W.; Xu, J.; Larsen, M.; et al. Targeting *Mycobacterium tuberculosis* tumor necrosis factor alpha-downregulating genes for the development of antituberculous vaccines. *MBio* **2016**, *7*, e01023-15. [CrossRef]
20. Sun, J.; Singh, V.; Lau, A.; Stokes, R.W.; Obregón-Henao, A.; Orme, I.M.; Wong, D.; Av-Gay, Y.; Hmama, Z. *Mycobacterium tuberculosis* nucleoside diphosphate kinase inactivates small GTPases leading to evasion of innate immunity. *PLoS Pathog.* **2013**, *9*, e1003499. [CrossRef]
21. Bhusal, R.P.; Bashiri, G.; Kwai, B.X.; Sperry, J.; Leung, I.K. Targeting isocitrate lyase for the treatment of latent tuberculosis. *Drug Discov. Today* **2017**, *22*, 1008–1016. [CrossRef]
22. Li, J.M.; Na, L.; Zhu, D.Y.; Wan, L.G.; He, Y.L.; Chun, Y. Isocitrate lyase from *Mycobacterium tuberculosis* promotes survival of *Mycobacterium smegmatis* within macrophage by suppressing cell apoptosis. *Chin. Med. J.* **2008**, *121*, 1114–1119. [CrossRef] [PubMed]
23. Meena, L.S. An overview to understand the role of PE_PGRS family proteins in *Mycobacterium tuberculosis* H37Rv and their potential as new drug targets. *Biotechnol. Appl. Biochem.* **2015**, *62*, 145–153. [CrossRef] [PubMed]
24. Deng, W.; Long, Q.; Zeng, J.; Li, P.; Yang, W.; Chen, X.; Xie, J. *Mycobacterium tuberculosis* PE_PGRS41 enhances the intracellular survival of *M. smegmatis* within macrophages via blocking innate immunity and inhibition of host defense. *Sci. Rep.* **2017**, *7*, 46716. [CrossRef] [PubMed]
25. Long, Q.; Xiang, X.; Yin, Q.; Li, S.; Yang, W.; Sun, H.; Xie, J.; Deng, W. PE_PGRS62 promotes the survival of *Mycobacterium smegmatis* within macrophages via disrupting ER stress-mediated apoptosis. *J. Cell. Physiol.* **2019**, *234*, 19774–19784. [CrossRef] [PubMed]
26. Yang, W.; Deng, W.; Zeng, J.; Ren, S.; Ali, M.K.; Gu, Y.; Li, Y.; Xie, J. *Mycobacterium tuberculosis* PE_PGRS18 enhances the intracellular survival of M. smegmatis via altering host macrophage cytokine profiling and attenuating the cell apoptosis. *Apoptosis* **2017**, *22*, 502–509. [CrossRef]
27. Danelishvili, L.; Yamazaki, Y.; Selker, J.; Bermudez, L.E. Secreted *Mycobacterium tuberculosis* Rv3654c and Rv3655c proteins participate in the suppression of macrophage apoptosis. *PLoS ONE* **2010**, *5*, e10474. [CrossRef]
28. Rengarajan, J.; Bloom, B.R.; Rubin, E.J. Genome-wide requirements for *Mycobacterium tuberculosis* adaptation and survival in macrophages. *Proc. Natl. Acad. Sci. USA* **2005**, *102*, 8327–8332. [CrossRef]
29. Zhang, W.; Lu, Q.; Dong, Y.; Yue, Y.; Xiong, S. Rv3033, as an Emerging Anti-apoptosis Factor, Facilitates Mycobacteria Survival via Inhibiting Macrophage Intrinsic Apoptosis. *Front. Immunol.* **2018**, *9*, 2136. [CrossRef]
30. Danelishvili, L.; Everman, J.; McNamara, M.; Bermudez, L. Inhibition of the plasma-membrane-associated serine protease cathepsin G by *Mycobacterium tuberculosis* Rv3364c suppresses caspase-1 and pyroptosis in macrophages. *Front. Microbiol.* **2012**, *2*, 281. [CrossRef]
31. Shin, D.M.; Jeon, B.Y.; Lee, H.M.; Jin, H.S.; Yuk, J.M.; Song, C.H.; Lee, S.H.; Lee, Z.W.; Cho, S.N.; Kim, J.M.; et al. *Mycobacterium tuberculosis* eis regulates autophagy, inflammation, and cell death through redox-dependent signaling. *PLoS Pathog.* **2010**, *6*, e1001230. [CrossRef]
32. Joseph, S.; Yuen, A.; Singh, V.; Hmama, Z. *Mycobacterium tuberculosis* Cpn60. 2 (GroEL2) blocks macrophage apoptosis via interaction with mitochondrial mortalin. *Biol. Open* **2017**, *6*, 481–488. [CrossRef] [PubMed]
33. Dutta, N.K.; Mehra, S.; Martinez, A.N.; Alvarez, X.; Renner, N.A.; Morici, L.A.; Pahar, B.; Maclean, A.G.; Lackner, A.A.; Kaushal, D. The stress-response factor SigH modulates the interaction between *Mycobacterium tuberculosis* and host phagocytes. *PLoS ONE* **2012**, *7*, e28958. [CrossRef]

34. Raman, S.; Song, T.; Puyang, X.; Bardarov, S.; Jacobs, W.R.; Husson, R.N. The alternative sigma factor SigH regulates major components of oxidative and heat stress responses in *Mycobacterium tuberculosis*. *J. Bacteriol.* **2001**, *183*, 6119–6125. [CrossRef] [PubMed]
35. Mehra, S.; Golden, N.A.; Stuckey, K.; Didier, P.J.; Doyle, L.A.; Russell-Lodrigue, K.E.; Sugimoto, C.; Hasegawa, A.; Sivasubramani, S.K.; Roy, C.J.; et al. The *Mycobacterium tuberculosis* stress response factor SigH is required for bacterial burden as well as immunopathology in primate lungs. *J. Infect. Dis.* **2012**, *205*, 1203–1213. [CrossRef] [PubMed]
36. Londono, C.; Osorio, C.; Gama, V.; Alzate, O. Mortalin, apoptosis, and neurodegeneration. *Biomolecules* **2012**, *2*, 143–164. [CrossRef]
37. Kaur, D.; Mathew, S.; Nair, C.; Begum, A.; Jainanarayan, A.K.; Sharma, M.; Brahmachari, S.K. Structure based drug discovery for designing leads for the non-toxic metabolic targets in multi drug resistant *Mycobacterium tuberculosis*. *J. Transl. Med.* **2017**, *15*, 261. [CrossRef]
38. Paik, S.; Choi, S.; Lee, K.I.; Back, Y.W.; Son, Y.J.; Jo, E.K.; Kim, H.J. *Mycobacterium tuberculosis* acyl carrier protein inhibits macrophage apoptotic death by modulating the reactive oxygen species/c-Jun N-terminal kinase pathway. *Microbes Infect.* **2019**, *21*, 40–49. [CrossRef]
39. Li, F.; Feng, L.; Jin, C.; Wu, X.; Fan, L.; Xiong, S.; Dong, Y. LpqT improves mycobacteria survival in macrophages by inhibiting TLR2 mediated inflammatory cytokine expression and cell apoptosis. *Tuberculosis* **2018**, *111*, 57–66. [CrossRef]
40. Su, H.; Zhu, S.; Zhu, L.; Huang, W.; Wang, H.; Zhang, Z.; Xu, Y. Recombinant lipoprotein Rv1016c derived from *Mycobacterium tuberculosis* is a TLR-2 ligand that induces macrophages apoptosis and inhibits MHC II antigen processing. *Front Cell. Infect. Microbiol.* **2016**, *6*, 147. [CrossRef]
41. Blasco, B.; Chen, J.M.; Hartkoorn, R.; Sala, C.; Uplekar, S.; Rougemont, J.; Pojer, F.; Cole, S.T. Virulence regulator EspR of *Mycobacterium tuberculosis* is a nucleoid-associated protein. *PLoS Pathog.* **2012**, *8*, e1002621. [CrossRef]
42. Wang, L.; Zuo, M.; Chen, H.; Liu, S.; Wu, X.; Cui, Z.; Yang, H.; Liu, H.; Ge, B. *Mycobacterium tuberculosis* lipoprotein MPT83 induces apoptosis of infected macrophages by activating the TLR2/p38/COX-2 signaling pathway. *J. Immunol.* **2017**, *198*, 4772–4780. [CrossRef] [PubMed]
43. Jin, C.; Wu, X.; Dong, C.; Li, F.; Fan, L.; Xiong, S.; Dong, Y. EspR promotes mycobacteria survival in macrophages by inhibiting MyD88 mediated inflammation and apoptosis. *Tuberculosis* **2019**, *116*, 22–31. [CrossRef] [PubMed]
44. Etna, M.P.; Giacomini, E.; Severa, M.; Coccia, E.M. Pro-and anti-inflammatory cytokines in tuberculosis: A two-edged sword in TB pathogenesis. *Semin. Immunol.* **2014**, *26*, 543–551. [CrossRef] [PubMed]
45. Balcewicz-Sablinska, M.K.; Gan, H.; Remold, H. Interleukin 10 produced by macrophages inoculated with *Mycobacterium avium* attenuates mycobacteria-induced apoptosis by reduction of TNF-α activity. *J. Infect Dis.* **1999**, *180*, 1230–1237. [CrossRef] [PubMed]
46. Feng, C.G.; Kullberg, M.C.; Jankovic, D.; Cheever, A.W.; Caspar, P.; Coffman, R.L.; Sher, A. Transgenic mice expressing human interleukin-10 in the antigen-presenting cell compartment show increased susceptibility to infection with *Mycobacterium avium* associated with decreased macrophage effector function and apoptosis. *Infect. Immun.* **2002**, *70*, 6672–6679. [CrossRef] [PubMed]
47. Patel, N.R.; Swan, K.; Li, X.; Tachado, S.D.; Koziel, H. Impaired *M. tuberculosis*—Mediated apoptosis in alveolar macrophages from HIV+ persons: Potential role of IL-10 and BCL-3. *J. Leukoc. Biol.* **2009**, *86*, 53–60. [CrossRef]
48. Pollara, G.; Turner, C.T.; Tomlinson, G.S.; Bell, L.C.; Khan, A.; Peralta, L.F.; Ricciardolo, F.L. Exaggerated in vivo IL-17 responses discriminate recall responses in active TB. *bioRxiv* **2019**. bioRxiv: 516690.
49. Cruz, A.; Ludovico, P.; Torrado, E.; Gama, J.B.; Sousa, J.; Gaifem, J.; Appelberg, R.; Rodrigues, F.; Cooper, A.M.; Pedrosa, J.; et al. IL-17A promotes intracellular growth of *Mycobacterium* by inhibiting apoptosis of infected macrophages. *Front. Immunol.* **2015**, *6*, 498. [CrossRef]
50. Deng, J.H.; Chen, H.Y.; Huang, C.; Yan, J.M.; Yin, Z.; Zhang, X.L.; Pan, Q. Accumulation of EBI3 induced by virulent *Mycobacterium tuberculosis* inhibits apoptosis in murine macrophages. *Pathog. Dis.* **2019**, *77*, ftz007. [CrossRef]
51. Zhou, X.; Li, X.; Wu, M. miRNAs reshape immunity and inflammatory responses in bacterial infection. *Signal Transduct. Target. Ther.* **2018**, *3*, 14. [CrossRef]

52. Liu, Y.; Jiang, J.; Wang, X.; Zhai, F.; Cheng, X. miR-582-5p is upregulated in patients with active tuberculosis and inhibits apoptosis of monocytes by targeting FOXO1. *PLoS ONE* **2013**, *8*, e78381. [CrossRef] [PubMed]
53. Huang, J.; Jiao, J.; Xu, W.; Zhao, H.; Zhang, C.; Shi, Y.; Xiao, Z. MiR-155 is upregulated in patients with active tuberculosis and inhibits apoptosis of monocytes by targeting FOXO3. *Mol. Med. Rep.* **2015**, *12*, 7102–7108. [CrossRef] [PubMed]
54. Rothchild, A.C.; Sissons, J.R.; Shafiani, S.; Plaisier, C.; Min, D.; Mai, D.; Gilchrist, M.; Peschon, J.; Larson, R.P.; Bergthaler, A.; et al. MiR-155–regulated molecular network orchestrates cell fate in the innate and adaptive immune response to *Mycobacterium tuberculosis*. *Proc. Natl. Acad. Sci. USA* **2016**, *113*, E6172–E6181. [CrossRef] [PubMed]
55. Xi, X.; Zhang, C.; Han, W.; Zhao, H.; Zhang, H.; Jiao, J. MicroRNA-223 is upregulated in active tuberculosis patients and inhibits apoptosis of macrophages by targeting FOXO3. *Genet. Test. Mol. Biomark.* **2015**, *19*, 650–656. [CrossRef]
56. Sharbati, J.; Lewin, A.; Kutz-Lohroff, B.; Kamal, E.; Einspanier, R.; Sharbati, S. Integrated microRNA-mRNA-analysis of human monocyte derived macrophages upon *Mycobacterium avium subsp. hominissuis* infection. *PLoS ONE* **2011**, *6*, e20258. [CrossRef]
57. Wang, Q.; Liu, S.; Tang, Y.; Liu, Q.; Yao, Y. MPT64 protein from *Mycobacterium tuberculosis* inhibits apoptosis of macrophages through NF-kB-miRNA21-Bcl-2 pathway. *PLoS ONE* **2014**, *9*, e100949. [CrossRef]
58. Xue, X.; Qiu, Y.; Yang, H.L. Immunoregulatory role of MicroRNA-21 in macrophages in response to bacillus calmette-guerin infection involves modulation of the TLR4/MyD88 signaling pathway. *Cell. Physiol. Biochem.* **2017**, *42*, 91–102. [CrossRef]
59. Zhao, Z.; Hao, J.; Li, X.; Chen, Y.; Qi, X. MiR-21-5p regulates mycobacterial survival and inflammatory responses by targeting Bcl-2 and TLR4 in *Mycobacterium tuberculosis*-infected macrophages. *FEBS Lett.* **2019**, *593*, 1326–1335. [CrossRef]
60. Zhang, G.; Liu, X.; Wang, W.; Cai, Y.; Li, S.; Chen, Q.; Liao, M.; Zhang, M.; Zeng, G.; Zhou, B.; et al. Down-regulation of miR-20a-5p triggers cell apoptosis to facilitate mycobacterial clearance through targeting JNK2 in human macrophages. *Cell Cycle* **2016**, *15*, 2527–2538. [CrossRef]
61. Zhang, D.; Yi, Z.; Fu, Y. Downregulation of miR-20b-5p facilitates *Mycobacterium tuberculosis* survival in RAW 264.7 macrophages via attenuating the cell apoptosis by Mcl-1 upregulation. *J. Cell. Biochem.* **2019**, *120*, 5889–5896. [CrossRef]
62. Tripathi, A.; Srivastava, V.; Singh, B.N. hsa-let-7b-5p facilitates *Mycobacterium tuberculosis* survival in THP-1 human macrophages by Fas downregulation. *FEMS Microbiol. Lett.* **2018**, *365*, fny040. [CrossRef] [PubMed]

MDPI
St. Alban-Anlage 66
4052 Basel
Switzerland
Tel. +41 61 683 77 34
Fax +41 61 302 89 18
www.mdpi.com

Pathogens Editorial Office
E-mail: pathogens@mdpi.com
www.mdpi.com/journal/pathogens

www.ingramcontent.com/pod-product-compliance
Lightning Source LLC
LaVergne TN
LVHW070636170726
843515LV00004B/261
* 9 7 8 3 0 3 9 3 6 6 5 8 3 *